QUEST FOR THE PILLAR OF GOLD
The Mines & Miners of the Grand Canyon

by

George H. Billingsley
U.S. Geological Survey
2255 N. Gemini Drive
Flagstaff, Arizona 86001

Earle E. Spamer
Academy of Natural Sciences
1900 Benjamin Franklin Parkway
Philadelphia, Pennsylvania 19103-1195

Dove Menkes
2530 Coventry Circle
Fullerton, California 92833

GRAND CANYON ASSOCIATION

P.O. Box 399
Grand Canyon, AZ 86023
www.thecanyon.com/gca

Grand Canyon Association is a not-for-profit organization. All proceeds from the sale of this book will be used to support the educational goals of Grand Canyon National Park.

Cover: The Grandview Mine, in a photo taken by Arion (John R. Putnam) around 1900. The Last Chance Mine dump can be seen in the lower right corner of the photo, near the bottom of the Redwall Limestone cliff. View is to the northwest from the top of the Redwall. Note also the surface structures and the two mule trains headed up the trail. Photo courtesy of Seaver Center for Western History Research, Natural History Museum of Los Angeles County, California, photo no. 8596.

© Copyright 1997 Grand Canyon Association

Monograph Number 10

ISBN 0-938216-56-2

LCN 96-080426

Editing by Sandra Scott and L. Greer Price

Design by Sullivan Scully Design Group/Mary Williams

Prepared in cooperation with the U.S. Geological Survey

Printed in the United States of America on recycled paper.

ACKNOWLEDGEMENTS

A publication of this scope is impossible to compile without the sincere interest of scores of people in libraries, historical societies, schools, and homes. Those who contributed their personal remembrances of people and events or their personal knowledge of places are by far our most valuable sources. Sadly, some of them have not lived to see this book.

We express sincere gratitude to Louise Hinchliffe and Valerie Meyer, former librarians of the Grand Canyon National Park Research Library; Katherine Bartlett and Dorothy House, former Museum of Northern Arizona librarians; John Irwin, Special Collections Librarian, Northern Arizona University, and Flagstaff City-Coconino County Public Library; Nancy Caudill, Librarian at California State University, Fullerton; Arthur L. Olivas, Photographic Archivist, Museum of New Mexico; Nancy Malone, Tom Jellinek, Lorna Kobayashi, Joanne Hardy, and staff, Fullerton Public Library; Catherine Ida and Sharlene LaForte, librarians at California State University, Long Beach; Norma Cochran, Librarian, Mohave County Historical Society; Blanche V. Berreman and Candy Owens, Coconino County Recorder's Office; Bonnie Greer, Arizona Historical Society, Northern Arizona Division, Pioneer Museum.

For their interesting historical accounts and hospitality, we extend special gratitude to Pat Bundy, J. Harvey Butchart, John and Merribeth Riffey, Jack and Marie Fuss, George Tanner, Leonard and Ruth Pemberton, and Reed Matthis.

Sincere thanks are also extended to Otis "Dock" Marston; Pat Hillard; Carol Patrick, Sue Abbie, and Michael Wurtz, of the Sharlot Hall Historical Society; Norma Bailey, Director of the Mohave County Historical Society; Susie Sato, Arizona Historical Foundation, Hayden Library, Arizona State University; Daniella Moneta, Richard Buchen and staff of the Southwest Museum; William Frank, Virginia Renner, Leona Schonfield, Fred Perez and staff of the Huntington Library; John Cahoon and Errol Stevens, Seaver Center, Los Angeles County Museum of Natural History; Peter Palmquist; P.T. Reilly for sharing his knowledge about Grand Canyon and its mines; Nellie I. Cox for permission to quote some passages from her two great books of the Arizona Strip country; Nick Mills of Boston University; Carol Downey and staff of the Arizona Department of Library, Archives and Public Records; Kathy Chase; Robert C. Euler; C.L. Sonnichsen, chief of publications of the Arizona Historical Association; Diane Baine and Ken Phillips, Arizona Department of Mines; Leroy Condie and Jolene Lindsay, National Park Service, Denver; and Dan Cassidy, Glen Dawson, Ellis Yochelson, Jim Ohlman, and the staff of the Yavapai County Recorder. Our thanks also go to Karen J. Wenrich, Hoyt Sutphin, Bradley S. Van Gosen, William L. Chenoweth, Marjorie E. MacLachlan, and Richard Quartaroli for reading and reviewing earlier material relating to this book. For recent information about the Bat Cave, we thank especially Kenton and Diane Grua, Richard Quartaroli, and Paul Martin.

The encouragement, suggestions, and advice of Susan Billingsley have proven invaluable. Additional financial support and encouragement by Karen J. Wenrich of the U.S. Geological Survey, Denver, Colorado, and Bonnie Murchey, U.S. Geological Survey, Menlo Park, California, is greatly appreciated.

Thanks to John Wunderlich for helping with word processing of early drafts.

For reading and commenting on the final manuscript, we thank Michael Anderson, Helen Fairley, Donaldson Koons, and L. Greer Price.

In the course of their infatuation with the Canyon, the authors have learned from many hikers, river runners, scholars, and field companions. This project evolved through a number of stages for about twenty years, and despite records, time may have taken its due from our ability to remember some of our sources. We apologize for the inadvertent omission of any of our kind contributors.

John Hauert Riffey
in a 1974 photo by
Art Pergam, Boulder City, Nevada

DEDICATION

John Hauert Riffey, or Riffey as the locals referred to him, became a legend in his own time—legendary in the National Park Service because he stayed at the Tuweep Ranger Station from October 1942 until his death on July 9, 1980; legendary to his many friends because he might well have been the most wonderful person they had ever met. His gentleness, knowledge, skill, and humor combined into a life style that was honorable and respected.

I first met John Riffey in 1963 when, following a hike to Lava Falls Rapid, my hiking companions and I stopped at the ranger station and were invited in to visit. Little did I know that John Riffey would become one of my closest friends and my adopted (to use his own phrase) great, great, great grandfather.

John's degree was in range management, and when he and his wife Laura settled into the Tuweep Ranger station at Grand Canyon National Monument his job was to make sure that over-grazing would not occur within the monument.

In the early days it was an all-day affair to drive from the ranger station to town—either St. George (110 miles) or Kanab, Utah (75 miles). It took so long because there were thirteen gates to open and close en route. With two people, the job wasn't too bad. A person alone had to stop the truck, open the gate, get back in and drive through, stop the truck, close the gate and then get back into the truck and head to the next gate. Today cattle guards keep the herds from roaming pasture to pasture and there are no gates to worry about. John was able to drive to Kanab in about one and one-half hours by 1979.

John maintained the airstrip that was near the ranger station, and eventually learned to fly his own airplane, which he named Pogo because in it he could hop all around the Canyon. When Laura became ill with cancer, Pogo was used to fly south across the Canyon to a hospital in Cottonwood, Arizona. Laura passed away and when I met John he was a widower.

John made several trips to Salt Lake City "to do research." Actually, he was courting a University of Utah graduate student whom he had met when she came to Tuweep. John had worked with one of her professors doing time-temperature studies and collecting a herbarium of plants representing the flora from the rim of Grand Canyon

to the Colorado River (Colorado River Mile 179). John and Merribeth were married on August 28, 1965, on John's fifty-fourth birthday. Merribeth, in the meantime, received her Ph.D. in botany and spent about half of each year at Tuweep and the other half teaching college in Bellingham, Washington. After a long illness, Merribeth died on April 15, 1993.

The Riffeys would open their home to strangers in a most warm and cordial manner, and the ranger station became a fairly popular place to visit. Although John had no children, he adopted a large family of young admirers. It was not uncommon to have fifteen to twenty people at the ranger station during Thanksgiving, New Year's, or Easter.

John created names for critters as well as equipment. For instance, all spiders were named Charlotte (from E.B. White's book, *Charlotte's Web*); George was the barometer; Matilda was the front-end-loading tractor; Sparky was the generator; Big Scratchy was the road grader; Digby was a badger; Misty, the ring-tailed cat; Lazarus, the lizard; and all foxes were named Reynard.

Once as I listened to the two-way radio at park head-quarters, the dispatcher asked John if he was lonely. He replied, "Not yet but I can see them peeking over the hills." I asked John about his ability to handle isolation. He told me, "I'm like a cat, I like people but I don't require them."

John knew that handling people with kindness and respect worked better than using rough tactics. One day while flying Pogo he spotted deer hunters who had strayed into the monument boundaries. He dropped them a note that said: "You are in the national park and hunting is illegal here. I will go back to the ranger station to get my pickup truck and will be out there in about two hours. If you are still there when I arrive, I'll have to give you a ticket." I'm sure that he would have given them one had they been there but they were gone and the job was done with a minimum of fuss.

Those of us were touched by John's wonderful personality and gentle ways were very lucky. To us, Tuweep will always be synonymous with John Riffey.

Art Gallenson

CONTENTS

LIST OF ILLUSTRATIONS

IMPORTANT NOTICES

All abandoned mines are dangerous. Do not enter them. Those that still are owned by private citizens and corporate entities require permission for entry. The Hance and Bass asbestos mines in Grand Canyon have been closed to all visitation by the National Park Service. The National Park Service and the Hopi Tribe, for sacred cultural reasons, have designated the salt deposits downstream from the Little Colorado River confluence off-limits to all but sanctioned American Indian visitors.

The removal or destruction of natural or historical artifacts in Grand Canyon National Park and other national and state lands is prohibited by law. Today, even removing items to take them to authorities is illegal, and it results in the loss of scientific and historical information. Until examined by professionals, the articles must remain in context, so that proper observations and conclusions can be made. Please leave things as you find them for future study and for others to enjoy. Take only photographs.

Overnight hiking and camping in Grand Canyon National Park are controlled by the National Park Service. Permits must be obtained from the Backcountry Office at Grand Canyon, usually far in advance. Other lands administered by federal, state, and American Indian agencies may require separate permits and may have additional regulations. Travelers should contact the appropriate agencies in advance. Only hikers experienced in desert climates and capable of using a topographic map should tackle Grand Canyon trails. Water, or access to it, is not dependable. Climatic variations in Grand Canyon are dramatic at any time of year. Flash floods can occur in any tributary, even when it is not raining in the Canyon itself.

INTRODUCTION

El Tovar, the grand old hotel perched on the Canyon's South Rim, broods over some of the most stunning scenery on earth. Carved in the dark-stained Oregon pine lintel over the north veranda, facing the Canyon, is the inscription "Dreams of Mountains as in Their Sleep." Grand Canyon is what great mountains likely would dream about. For people, it brings musings of things that might be. We all have dreamed of wealth; and some men have gone deep in the Grand Canyon of the Colorado in search of it. Gold was their dream, yet the Canyon yielded almost none, and offered other things instead and reminders that wealth, and what we make of it, is relative.

American Indians were the earliest miners in Grand Canyon. Salt, red clay, and copper stones were valuable commodities for trading, healing, spiritual, and domestic uses. They traded information on the locations of mineral deposits to the early prospectors in return for European-American goods.

Income-producing metals and minerals—copper, lead, and asbestos mostly—were found in Grand Canyon by the late 1800s and early 1900s. Many of the deposits were small and not rich enough to make it worth the trouble and time to get them out of the canyon and to distant markets. A few of the mines were profitable, but they were of limited extent and life. Still, this did not deter itinerant prospectors, who usually worked alone but showed up in waves when gold was discovered. After all, prospectors "are eternally on the verge of untold riches" (Verkamp, 1940, 1993:14).

The history of mining in the Grand Canyon region has been eclipsed by the rich mineral resources and the colorful cultural history of mining in southern and central Arizona, New Mexico, Nevada, Utah, and California. No single synthesis of mining in the Canyon region has been prepared before, probably because this region has not been popularly associated with mining activity.

All but one of the mines within the physiographic boundaries of Grand Canyon are abandoned and are now part of Grand Canyon National Park or are on other federal or American Indian lands. The Hance asbestos mines are the sole exception.

The search for uranium in the Grand Canyon region, begun in the late 1950s, did not reach a feverish peak until the mid-1970s, when its price escalated. With the increase, geological phenomena called breccia pipes, in which most of the uranium ore in this region is found, were intensively studied, and exploration techniques improved. In the early 1980s uranium prices plummeted, and legal and political uncertainties, the reduction of nuclear arsenals, foreign competition, and the low market price have contributed to a declining trend in uranium exploration. As profitable mines exhaust their ores, operations cease altogether.

Perhaps the most valuable commodity Grand Canyon offers is also one of its scarcer ones—water. Without it, prospecting in the Canyon would be impossible. More money has been spent on water-related problems in the southwestern United States than any other mineral. (Indeed, water itself is a mineral. It has a definite chemical structure, and at lower temperatures it exhibits the crystalline structure we call a snowflake.) The human need is clear, but in many cases, too, the mining operation itself uses water for the extraction and initial processing of ores. In the days when pack animals and wagon haulers were needed, even more water was required. In an arid area with an expanding population, the development of Colorado River water—the only major, reliable water source in the American Southwest—has cost thousands of times more than all the mineral wealth that has ever been produced from the Grand Canyon region. No doubt, water will continue to be the major negotiable commodity in the region for years to come.

This Publication

This is a synopsis of the prospectors, their times and mines, and the minerals they found in and around the Grand Canyon. This is not an interpretive narration. It is the first consolidation of far-flung data on the subject, a starting point for more focused historical treatments. The area encompassed is about fifteen thousand square miles of northwestern Arizona, including what today are Grand Canyon National Park and parts of adjacent Kaibab National Forest, Glen Canyon National Recreation Area, Lake Mead National Recreation Area, Hualapai Indian Reservation, Havasupai Indian Reservation, Navajo Indian Reservation, rangelands administered by the U.S. Bureau of Land Management and the State of Arizona, and private lands (Map 1). Previous surveys of mining in the Grand Canyon region are restricted to brief, geographically restricted summaries (as cited throughout this publication), Billingsley's (1974) earlier review, and, for the western Grand Canyon region, an unpublished administrative report to the National Park Service (Foord et al, 1978).

We chose to arrange this book in a geographical presentation, from west to east across the region. The mining districts of this region (Map 2) have never been wholly formalized; because of this we have delineated them only approximately. In many instances, our information is published for the first time. The chronological series of events and mineralogical data by district and by mine have been synthesized in two appendices.

Topically, we could have limited ourselves to recounting mine production data and dates of claims, but many of these records are sketchy. More important to the story is the "frontier" spirit of all who came to the Canyon, from the first immigrants thousands of years ago, to the "pioneers" from the East. The history of human endeavors here is both rich and inseparable from the raw data of numbers and dates, even though many parts of the history are lost. With the interwoven lives of the American Indian peoples and the campaigns of European-American colonists, we are obliged also to pay attention to social, political, and legal concerns in the history of Grand Canyon mining. There are dozens more stories and anecdotes that we could have interjected into this publication, which simply will not fit, so as an introduction to the broader history of people at Grand Canyon, we direct the reader to three different perspectives: Verkamp (1940, first published in 1993), who itemized the events and people from the view of the pioneer and resident; Hughes (1978), who brought together the rich resources of historians and archivists to tell the story of American Indians, later adventurers, families, and the Canyon; and Pyne (1982), who looked at the Canyon from the perspective of scholars and explorers.

For many of the sites mentioned in this book some very specific locations are known to National Park Service and other supervising authorities. Provisions of the National Historic Preservation Act of 1966 (as amended through 1992) and the Archaeological Resources Protection Act of 1979 prohibit public disclosure of the nature and location of archaeological and historical resources when special permits are required for excavations, removal, or disturbance of artifacts, or when there is a "substantial risk of harm, theft, or destruction" to the resource or the area in which they are found. There are four disclosure classes, ranging from "readily disclosed to the public" to "officially closed to all visitation." This book makes mention of sites from all four classes. It has been reviewed by National Park Service personnel, and we have modified the text to accommodate the laws as well as some specific requests made by the reviewers. We believe that nondisclosure of specific locations of some of the sites does not affect the historical interpretations that can be made from the information we provide. Regarding geological terminology and data, earlier drafts of the book have been reviewed and approved by the U.S. Geological Survey.

Our book begins with a brief overview of the cultural and geographic setting of Grand Canyon. Next, we present a primer on the geology of Grand Canyon in order to establish a general framework for the discussions of geologic strata and structures in this region, which are inescapable when discussing Grand Canyon's mines. The main part of the publication deals with the history of mining in the region. Much of the information has until now been scattered through published literature, archival holdings, and ponderous county record books. Some oral traditions have been captured here. In many cases, the miners are of greater interest than their meager mines.

At no time before has the Canyon been accorded so many legal and jurisdictional safeguards than those that are in place today. The National Park Service is charged with balancing the preservation and enjoyment of everything within park boundaries. For better or worse, federal and state regulations and guidelines place controls on development, monitor environmental impacts, and encourage moderated recreational uses of the Canyon's widespread, fragile resources. Mining activity in the park is unthinkable now, but surrounding lands can expect various levels of economic development. The historic mining resources of the Canyon, so long ignored, need to be documented before more information is lost. The park service needs data to help formulate plans for recording and preserving the condition and relics of the mines; in some cases, to document as much as possible before closing off dangerous sites from visitation. This publication is hardly comprehensive enough to professionally document the contents of the hundreds of mines and claims in this region, but we hope that this is a useful complement to historical research and for the park service's record of the patchy history of miners and their work in and around Grand Canyon. We hope it is a stimulus to fully document the historical resources scattered throughout the region, before they are lost forever.

Protection of Grand Canyon

President Theodore Roosevelt, in a very short speech at Grand Canyon in 1903, charged the American people with the protection of Grand Canyon (Holt and Ferleger, 1941). He appealed:

> *Leave it as it is. You cannot improve on it. The ages have been at work on it, and man can only mar it. What you can do is keep it for your children, your children's children and for all who come after you, as one of the great sights which every American, if he can travel at all, should see. Keep the Grand Canyon as it is.*

What may have been in the back of the president's mind was something that was then recent history, and that could have changed the face of Grand Canyon forever—the Denver, Colorado Cañon, and Pacific Railroad.

Unlike many development projects of the day, at a time when the nation was growing with wonderful new tools—electricity, telephones, automobiles, and (in a few months) airplanes—there was no general cry for a railroad along the Colorado River. It was still a remote region, and many of those who knew of it saw no need to go there. The great mining fields of central and southern Arizona, Nevada, California, and elsewhere in the West, eclipsed the meager workings of the Grand Canyon region. Still, the Canyon and its river represented a major thoroughfare through the western deserts and the Rocky Mountains.

To railroad and coal barons, an easy grade through the heart of the canyonlands had a strong appeal. A river-grade railroad from the Rocky Mountains to the Pacific Ocean would bring coal quickly and efficiently to the West Coast and ports there, and carry the mineral wealth of California and the Southwest to railroads linking eastern markets.

To establish the engineering feasibility of a railroad from Denver through Utah and Arizona to San Diego, the Denver, Colorado Cañon, and Pacific Railroad was established, and Robert Brewster Stanton set out down the Green and Colorado Rivers in 1889 to survey the route (Stanton, 1965; Smith and Crampton, 1987). His party met with disaster in Marble Canyon, where three men, including Frank Mason Brown, president of the railroad company, drowned. Stanton regrouped and the survey was completed in 1890. The route was a feasible one. In fact, it was deemed even more easy to build than some routes that had already been put through the Rocky Mountains. The plan was not without its detractors (see, for example, Lenz 1986:79), who claimed knowledge of the impractical aspects of the route. In any case, the capital needed to build the railroad was never gathered. Vast oil reserves, discovered soon thereafter, changed the economy of energy production.

Today it is our good fortune that the railroad was never built. Not only would the aesthetics of the inner canyon have been wrecked with roadbed, trestles, tunnels, and even a switchyard (planned for the open area north of Bass Rapid), but also by the trappings of civilization that followed all of the new western railroads. This was the railroad's plan, after all. It established the Colorado Grand Canyon Mining and Improvement Company in Denver, whose "stated purposes were: the operation of mines along the Canyon, located by the Stanton expedition, the location of new claims, the establishment of toll roads and a ferry across the river, the operation of hotels on the rim, and the acquisition and sale of real estate" (Verkamp, 1940, 1993:17, citing Record of Articles of Incorporation, Coconino County, Territory of Arizona, vol. 1, pp. 8-9, Recorder's office, Flagstaff). Stanton himself envisioned "… each cove with its picturesque Swiss chalet, and its happy mountain people with their herds of sheep and mountain goats, developing local business for our future railroad" (Stanton, 1965:152).

His daydream was a bit ethereal, but imagine how the railroad would have affected Grand Canyon's mines. Haulage costs would have been dramatically reduced, lower grade ores would have been economically viable, and there would have been virtually no limit on the amount of ore—and quarrystone—that could have been taken from the Canyon. Mining camps may have sprung up, perhaps even towns, their people and ore-processing plants watered by the Colorado River, sucked up by pumps driven by electricity from turbines powered with its own flow. Perhaps the Canyon, or most of it, would have been made a national park anyway, but it would have been less grand and certainly less remote.

Grand Canyon National Park

The early history of mining in Grand Canyon is hardly a success story. Only a few people made modest gains. Once transportation became less an obstacle, the probability of future profits faded as the United States government took progressively stronger interest in making Grand Canyon the property of the American people—despite the men who wanted to own it.

In 1882, Senator Benjamin Harrison of Indiana introduced the first bill in Congress to establish Grand Canyon as a national park, but the bill never went to a vote. As president in 1893 Harrison proclaimed the eastern part of the Canyon and the surrounding area as the Grand Canyon Forest Reserve, exempting it from homestead laws and all other public land laws except those involving mineral claims.

During the early 1900s, Congress began to move to protect the Canyon, or to at least get politically involved in the process. On November 28, 1906, President Theodore Roosevelt signed a bill establishing the Grand Canyon Game Preserve. This action protected the deer and other game animals, but not the predators. A U.S. Attorney opinion also closed the preserve to all private entry. Roosevelt built a cabin at Ryan, near Warm Springs Canyon, and hunted mountain lions. (So popular was the sport that within ten years mountain lions were almost gone from the plateau, and the Kaibab deer herd, then without its primary predator, grew to more than fifty thousand. Overpopulation lead to disease and starvation—a classic illustration of natural balance.)

On January 11, 1908, Roosevelt, in a controversial move invoking the Act for the Preservation of American Antiquities which required no congressional checks or approvals, established Grand Canyon National Monument. The primary directive of this proclamation was to forbid prospecting and mining on all lands in the eastern part of Grand Canyon that were not already covered by valid claims. Most of the mines were abandoned by this time anyway, and those that were allowed to continue faced increasing haulage costs and lower ore prices that forced most of the small operations out of business by the 1920s.

On February 26, 1919, President Woodrow Wilson signed the bill that created Grand Canyon National Park. Congress revised the boundaries of the park in 1927, eliminating all future mining and other developments within them. A new Grand Canyon National Monument was proclaimed by President Herbert C. Hoover on December 22, 1932, which included many tributary canyons to the Colorado River as well as central parts of the North and South Rims (from about Colorado River Mile 145 to 184). Efforts to acquire small inholdings of private or state-owned land by gift or exchange added many mining claims (Hughes, 1967). By

1975, John Hance's old asbestos claims in Asbestos Canyon were the last mining claims that had not reverted to the National Park Service. The 312.42-acre parcel is owned by the descendants of William Randolph Hearst and since the 1980s negotiations for its transfer to the National Park Service have been underway (John Ray, Grand Canyon National Park, personal communication, 1988).

Significant expansion of Grand Canyon National Park boundaries in 1975 protected against future mining virtually all of the tributary canyons within the physiographic bounds of Grand Canyon. The only excluded areas are the Parashant and Whitmore Canyons, where in 1975 uranium exploration was conducted by Exxon Minerals, USA. Adjacent to the national park are segments of the Hualapai, Havasupai, and Navajo Indian Reservations, and lands administered by the forest service and the Bureau of Land Management. Mining on the Indian reservations remains under the jurisdiction of tribal laws. The boundaries along the Colorado River between the national park and the Navajo Indian Reservation (on the east side of Marble Canyon) and Hualapai Indian Reservation/Grand Canyon National Park boundary (between River Miles 165 and 273) remain in dispute, taken variously to be the canyon rim, the historic Colorado River high-water line, or the center of the river. (The Havasupai Indian Reservation boundary does not reach the Colorado River.)

Cultural and Political Boundaries

Small farming or ranching communities near the Canyon were founded as early as the 1860s. North of Grand Canyon, Mount Trumbull (Bundyville) and Pipe Springs, Arizona, and Kanab and St. George, Utah, were the settlements nearest to the Canyon. From these communities prospectors began their long treks (about seventy miles from St. George) southward to Grand Canyon and, they hoped, riches. The river gateways to the Canyon, Lees Ferry upstream (Mile 0), and Pearce Ferry downstream (Mile 279.5), provided the only Colorado River crossings in the region. The river itself remained virtually impassable.

South of the Canyon, Peach Springs, Seligman, Ashfork, Williams, and Flagstaff, Arizona, were established in the early 1880s during the construction of the Atlantic and Pacific Railroad. Each town lies about seventy-five miles south of Grand Canyon, except Peach Springs, which is twenty-one miles by Peach Springs Canyon to the Colorado River. That route remains the easiest and most accessible to the bottom of the Canyon.

Grand Canyon Village was established on the South Rim by the late 1890s. It began as a cluster of small hotels and tents. The stage lines that reached it also serviced outlying hotel-camps run by entrepreneurial prospectors. In 1900, a railroad spur was built from Williams to Anita, a mine about twelve miles south of Grand Canyon Village, and was extended to the village in 1901.

Lack of water is a recurring theme in this history of resources exploitation in the Grand Canyon region. Vast deserts lie to the east and west of Grand Canyon. Food and water for animals was nearly nonexistent, so even travel through, much less settlement of these areas was difficult before the invention of the automobile.

Grand Canyon National Park, as of 1994, encompasses 1,218,376 acres (1,904 square miles) of rugged landscape, taking in not only much of the physiographic canyon, but also adjacent portions of the plateaus into which the chasm has been cut. Large parts of the physiographic Grand Canyon are not within the national park, however. Parts lie within the Hualapai Indian Reservation (western Grand Canyon), the Havasupai Indian Reservation (central Grand Canyon), and the Navajo Indian Reservation (eastern Grand Canyon). Several large tributaries lie within federal lands administered by the U.S. Forest Service and the U.S. Bureau of Land Management. The few privately owned lands within the walls of the Canyon are located north of the Colorado River in Whitmore and Toroweap Canyons (western Grand Canyon), and one small parcel is near the bottom of the canyon in its eastern part.

Geographic Setting

The Canyon lies in the southwestern part of the Colorado Plateau, a broad, segmented series of tablelands drained by the Colorado River and its principal tributary, the Green River. It is an area of locally folded, mostly flat-lying strata of Paleozoic and Mesozoic ages with an average regional dip of one to two degrees to the northeast. Where the river exits the Colorado Plateau at the Grand Wash Cliffs, the physiographic boundary is pronounced and dramatic. To the west of the high cliffs at lower altitude is a region of strongly faulted terrain of mountains and basins.

Grand Canyon is a complex of tributary canyons, each with its own drainage area, contributing to the Colorado River. The largest tributaries are the Little Colorado River, Diamond Creek, and the creeks of Parashant, Kanab, and Havasu Canyons.

Local physiographic areas or subprovinces within the Colorado Plateau were described by early geologists and geographers on the basis of their physical characteristics, similar structure, climate, and a unified geomorphic history. Near Grand Canyon, each subprovince has its own geomorphic or physiographic characteristics with local structure or topographical boundaries that make them well defined. However, several miles away from the Canyon, some boundaries become vague and ill defined due to the deterioration of the characters that make them stand out so well near the Canyon. In Map 1, these indistinct boundaries are indicated by long dashed lines, and they are drawn on the map where the terrain is most typical of the original physiographic description (Billingsley and Hendricks, 1989).

The physiographic expression of Grand Canyon itself limits access to its depths, but early American Indian residents and later European-American prospectors found many usable routes into the chasm. Some of these routes were upgraded to trails which could accommodate pack animals, but even most of these subsequently were abandoned. Hikers today find these trails challenging, and readers are here cautioned to the hazards of Grand Canyon hiking. Many useful guides are available (see listings in Spamer, 1990c, 1993b, and Thybony, 1994).

The Colorado River flows east to west through Grand Canyon. By convention, distances along it are measured downstream from Mile 0 at Lees Ferry, Arizona. There, at the end of Glen Canyon and the beginning of Marble Canyon, is also the eastern boundary of Grand Canyon National Park and the stratigraphic top of Grand Canyon's rock formations, the Kaibab Formation.

The arbitrary end of Marble Canyon is where Nankoweap Creek enters the Colorado River, near Mile 52. There, the physiographic appearance of the chasm begins to change, opening up into Grand Canyon. Some river runners consider Marble Canyon to continue to the confluence with the Little Colorado River (Mile 61). Indeed, along the river itself the perspective precludes seeing the grand opening up of the canyon until after the Little Colorado River is passed.

For descriptive purposes, Royal Arch Creek (Mile 116) on the south, and the Powell Plateau on the north, delineate the margin between an open eastern Grand Canyon and the T-shaped profile of the central Grand Canyon.

At Mile 180, at Prospect Canyon (on the south side) and Toroweap Canyon (north side), another arbitrary boundary is drawn between McKee Point (south) and Toroweap Point (north), demarcating the central from the western Grand Canyon. Physiographically, the Canyon to

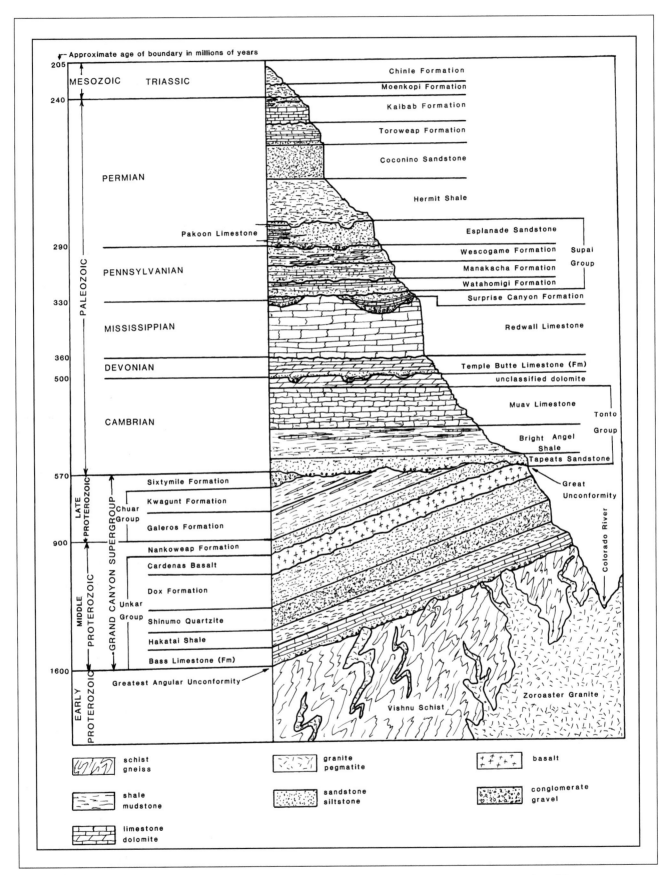

Figure 1. Schematic cross-section of the Grand Canyon.
After Billingsley & Hendricks, 1989:67-71.

the west becomes more open and wider, with inner canyon plateaus, mesas, and large landslide blocks.

The western portal to Grand Canyon, at Mile 277, is along the Grand Wash Cliffs that tower over artificially created Lake Mead, the impoundment behind Hoover Dam (formerly Boulder Dam). The natural features of the Colorado River corridor have been flooded by Lake Mead upstream to about Mile 239, Separation Canyon.

Geologic Setting

Geologists have differing opinions on the actual amount of time required to excavate the Canyon, ranging from about four to five million years. The weathering processes of water, wind, and naturally occurring organic chemicals have slowly eroded the rocks of the Colorado Plateau. The debris, falling into the Canyon as boulders and cobbles, is constantly moved down through tributary canyons and ground into finer fragments, which eventually are dumped into the Colorado River. Until the great period of dam construction on the Colorado River in the middle part of the twentieth century, the river, also working away at the debris, finally delivered sand and silt to the Gulf of California. Since 1936, when the gates of Hoover Dam were closed, virtually all of the sediment excavated from Grand Canyon and its tributary canyons has been impounded in Lake Mead on the Arizona-Nevada border. Clearer water passes on to the gulf.

Figure 1 provides a schematic diagram of the names and positions of the geological strata that typically crop out in Grand Canyon. For more details of the geology of Grand Canyon, Elston et al. (1989) and Beus and Morales (1990) are recommended technical publications. Annotated bibliographies of Grand Canyon geology, reviews of geological studies in Grand Canyon, and extensive compilations of Grand Canyon geological data have been compiled by Spamer (1983-1993a).

The Canyon's lowermost and oldest rocks began as sedimentary and volcanic rocks that were subjected to intense heat and deformation about 1.7 billion years ago (Early Proterozoic Era). These rocks encompass all different kinds of metamorphic, igneous, and granitic rocks, such as schists, gneisses, and granites.

The schists and granites, found mostly in the eastern and western Canyon, were formed into a mountainous terrain and subjected to erosion for about 500 million years. This long period of erosion wore the mountains to a nearly flat plain before the next accumulation of sedimentary muds, sands, and limemuds were deposited during the Middle Proterozoic Era, about 1.2 billion years ago, and lasted for about 300 million years into the Late Proterozoic Era. The unconformable break between the Early and Middle Proterozoic is called the greatest angular unconformity (Figure 1). The sediments of the Middle and Late Proterozoic, together called Grand Canyon Supergroup, accumulated to a thickness of about twelve thousand feet. These strata, now faulted and tilted so that they are no longer an uninterrupted pile, crop out in parts of the eastern and central Grand Canyon.

About 900 million years ago, huge blocks of Grand Canyon Supergroup rock strata were lifted upward, faulted, and tilted to create mountains and intervening valleys. Over a period of 330 million years, this block-faulted terrain was subsequently eroded to a lowland of small hills and wide valleys. The blocks of Late Proterozoic strata that were faulted downward were protected from erosion before the next sequence of sediments were deposited over them during the Paleozoic Era.

Beginning about 570 million years ago and lasting for about 500 million years (Paleozoic and Mesozoic Eras), the Grand Canyon region was subjected to periods of sediment accumulation, interspersed with major and minor periods of erosion. The horizon where the first flat-lying Paleozoic strata (usually the Tapeats Sandstone or the Bright Angel Shale; Figure 1) overlie the older schists, granites, or tilted strata of the Grand Canyon Supergroup is called the Great Unconformity. The erosional time gap represents about 830 million years between the Tapeats Sandstone/Bright Angel Shale and the schists and granites, and about 330 million years where the Tapeats/Bright Angel rests atop strata of the Grand Canyon Supergroup.

The sedimentary rocks of the Paleozoic Era make up most of the walls of Grand Canyon to the rim; the period of sediment accumulation and erosion represents about 325 million years. A short period of erosion lasted for about ten to thirty million years before more sediments were deposited during the Mesozoic Era. The Mesozoic strata once covered most, if not all, of the Grand Canyon region, in layers as thick as two thousand feet, but at Grand Canyon almost all of the Mesozoic rocks have been removed by Cenozoic Era erosion during the last sixty-five to seventy million years. Mesozoic rocks can be found nearest to the Canyon at the Vermilion Cliffs, Kaibito and Paria Plateaus, and small isolated remnant landmarks such as Shinumo Altar, Gold Hill, Cedar Mountain, Red Butte, Mount Trumbull, Mount Logan, Poverty Mountain, Poverty Knoll, and Diamond Butte (Map 1).

Breccia Pipes

Breccia pipes are special geological features with which many of the Canyon's mines in the Paleozoic strata are associated. Most early prospectors probably did not know the structure of these pipes (in fact, their true nature was not recognized even by geologists until about 1950) but they understood that the deposits were of limited extent. A description of the Grand Gulch Mine in the Mohave County Miner (July 25, 1914) observed that "The ore bodies occur around the sides of a vertical plug-like mass of rock, which is sedimentary but is entirely different from the stratified rocks that enclose it and will in this paper be called the filling." Figure 2 illustrates the general structure of a breccia pipe. The reader should bear in mind that in the Paleozoic rocks of northern Arizona they are found only between the Redwall Limestone and the plateau surface.

What is a breccia pipe? Simply, a pipe in this region begins as a cave in the Redwall Limestone. The cave is eroded by chemical dissolution during ground water seepage through cracks in the limestone, eventually enlarging it further. As a cave becomes larger, it reaches a size beyond which it is unable to support the roof. The roof begins to slowly collapse, or cave in, over a long period of time. As the cave ceiling breaks away in stages, roof collapses continue upward into overlying rocks until there is no more room for the collapsed material to accumulate (Figure 3). Ground water circulation removes some of the smaller particles from the debris, transporting them through the interconnecting cave system. This provides more room for the pile of broken rock to settle, which in turn allows more room near the ceiling for further collapse. The accumulation of breccia (broken rock) extends upward in a circular pipe-like column, hence the name breccia pipe (Figure 4).

Mineral waters circulating through the breccia open more spaces. The dissolved minerals in the water precipitate out in the open pore spaces, eventually cementing most of the breccia in an ore-rich matrix, most often abundant in copper, silver, and uranium. For reasons that are still not understood, many breccia pipes are not

mineralized. Less than 5 percent of all breccia pipes contain economical mineral deposits, and those that do appear to be scattered randomly.

The timing of the mineralization process in breccia pipes is ancient, dating to a period when Mesozoic strata overlay this part of the Colorado Plateau, and mountains were to the southwest in central Arizona. Analyses of the radiometric ages of uranium-mineral deposits from several breccia pipes indicate that there were two major periods of mineral emplacement, one about 260 million years ago, and another period 200 ± 20 million years ago (Ludwig and Simmons, 1992). The mode of emplacement is thought to have been by ground water leaching uranium from volcanic ashes deposited in the Chinle Formation, mobilized by a change

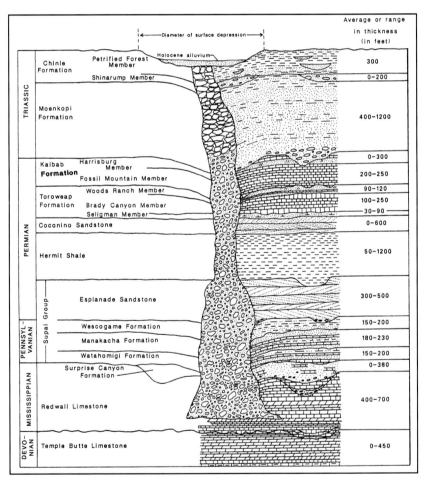

Figure 2. Schematic cross-section of a breccia pipe based on exposures in cliffs of the Grand Canyon. The unit thicknesses shown for the Triassic Chinle and Moenkopi strata represent their thickness ranges in the Grand Canyon area. The unit thicknesses for the lower Paleozoic strata correspond to thicknesses as seen in the Coconino Plateau of the eastern Hualapai Indian Reservation. After Van Gosen & Wenrich, 1989:5.

in hydrologic gradients caused by tectonic uplift of the region southwest of the Canyon. The fluids may have interacted with Proterozoic basement rocks before circulating into the breccia pipes.

Erosion of Grand Canyon during Cenozoic time has exposed several mineral rich breccia pipes (Figure 1). Thus, most of the mine locations in the Paleozoic rocks are found in the upper half of the walls of Grand Canyon.

Minerals of Grand Canyon

The mineral deposits mined in Grand Canyon are generally located in fault zones or intrusive rocks of the Early, Middle, and Late Proterozoic rocks; in breccia pipes of the upper Paleozoic rocks of the Redwall Limestone and higher; and in brecciated limestone beds of the Kaibab Formation on the Coconino, Kaibab, and Uinkaret Plateaus. There are no minable mineral deposits in the Cambrian or Devonian strata other than the salt found in the Cambrian Tapeats Sandstone.

Silver and gold were among the first metals that attracted prospectors to the Grand Canyon region. Some silver was found in Havasu Canyon in the early 1870s, but apparently

Figure 3. The cave in the lower part of the photo (arrow) is in the Redwall Limestone. It is filled with collapse breccia (A), allowing the overlying strata to bend inward and collapse. Photo is north of the Colorado River at Mile 270. Photo by George Billingsley, 1976.

did not capture the interest of prospectors on the Colorado Plateau as much as had the discovery of silver in the red sandstones north of St. George, Utah. The most widely known mine operations of that time were the Silver Reef mines in Utah (Crampton, 1972). When silver was not found in abundance, copper became the number one commodity sought in the Canyon, although visions of gold veins were probably always in the minds and dreams of early prospectors. Asbestos was discovered and mined from the canyon, too, but it was not widely developed, probably because the markets were too distant. In most cases, minerals were mined because they were there. It was better than nothing.

During the late 1950s, uranium was discovered in the Grand Canyon region, and it has remained the principal minable material to the present time, although its demand too is now on the decline. The high-grade uranium ores of

Figure 4. Bat Cave breccia pipe. The column of broken rock is 400 feet high, the result of a collapsed cave originating in the Redwall Limestone below the view seen in the photo. View is looking east about one mile southeast of the Bat Tower Viewpoint ("Guano Point") on the Hualapai Indian Reservation in western Grand Canyon (Grand Canyon West). Photo by George Billingsley, 1985.

the Canyon region made the development of mines here a profitable, if short-lived, venture despite declining prices through the 1980s. Silver is a rather high concentrate in some mineralized breccia pipes but it is not processed with the uranium ore because the mills that process the uranium are not set up for expensive silver recovery processes. The uranium mining activities and their environmental contamination problems have spurred additional measures for the protection of the Grand Canyon region, resulting in many legal and economic conflicts.

Appendix 2 is a list of metals and minerals found in the Grand Canyon region, mostly in the breccia pipes. The suite of minerals that have been detected contain just a few that are of economic value. Copper and uranium oxides are the principal ores of monetary concern.

MINING DISTRICTS OF
GRAND CANYON & VICINITY

Mining districts were often established by miners and prospectors whenever and wherever a group could get together. Several districts were established on the basis of the known extent of all claims in a specific area. Often, the boundaries were not known or were not very well specified. As a result, some overlapped those of other districts. Some prospectors made up their own mining district, perhaps not knowing or caring that their area lay within an already established district.

We delineate and describe the twelve known mining districts of the Grand Canyon region. Many of these districts are here redefined and elaborated upon for the purposes of identification and to aid discussions made in this publication. To our knowledge, no one has attempted to consolidate and define these data before. Garbani's (1993) compilation of Arizona mines and mining districts does include several districts in the Grand Canyon region, but the maps depict only some of the areas in which mining activity took place. They do not delineate contiguous districts, and they omit some of the informal districts that had been established by consortia of prospectors.

Our interpretations and findings, of course, are not intended to serve as any kind of legal document, nor could they. Rather, they are historical interpretations of a poorly explored aspect of human intervention in Grand Canyon. Parts of the information we present very likely will change as new data are discovered, collated, and compared with existing information. If mining activities should continue in the region, then naturally this publication will serve as a historical synopsis reaching almost the end of the twentieth century.

Bentley Mining District

Mining claims in a district organized soon after the discovery of copper on the Sanup Plateau (Grand Gulch Bench) along the Grand Wash Cliffs were relocated from Washington County, Utah, to Mohave, Arizona. On June 23, 1873, several miners formed the Bentley Mining District, named after Richard Bentley, who claimed the first copper in the area. The district boundaries were not well known at that time. Mining claims around Pierce Canyon, the Grand Gulch Mine area, the Hidden Canyon area to the north, and the Snyder Mine to the east, were filed as part of this district. To be consistent with geographic areas and to avoid confusion in later discussions, we omit the Snyder Mine area from the Bentley Mining District.

Lost Basin Mining District

Mining in this district is restricted mostly to gold placer claims along the Colorado River, in the Hualapai Wash area, and near the community of Meadview, Arizona. The eastern boundary of this district abuts the Music Mountain Mining District at the extreme northwestern part of the Hualapai Plateau along the present boundary between the Hualapai Indian Reservation and Grand Canyon National Park.

Music Mountain Mining District

A few isolated gold mines were claimed along the Grand Wash Cliffs southwest of the Hualapai Plateau, producing a brief flurry of gold mining activity, and mines were recorded during the 1880s at Kingman, Arizona, as located in the Music Mountain Mining District. Again, there were no definite boundaries for the district except that the area was roughly twenty-five miles north of Hackberry, Arizona, in the foothills of the Music Mountains. These mines, and a few placer claims in Spencer Canyon, a southern tributary to the Colorado River in the Hualapai Plateau area, are the only known claims in this district.

Centennial Mining District

This district came into existence in the late 1880s, centered around the Copper House Mine in the Andrus Canyon area north and west of the Colorado River at the mouth of Parashant Canyon (Mile 198.5). The western boundary has been included in the Bentley Mining District since the 1950s to accommodate the Snyder Mine area. For location purposes, geographic description, and centralization, the name Centennial Mining District will be maintained as originally named, but the center of activity will be relocated from Andrus Canyon to the Snyder Mine area because Andrus Canyon is very close to the older Copper House Mine, which is the center of the Mount Trumbull Mining District.

Mount Trumbull Mining District

Like most of the other mining districts, the Mount Trumbull district was established, with no defined boundaries, around a cluster of mines. The mines that began this district were located in Parashant Canyon during the late 1870s. Other mines and prospects, located farther to the east in the areas of Whitmore Canyon, Toroweap Valley, and Tuckup Canyon, did not specify a mining district. Since these areas are all within a reasonable distance of Mount Trumbull, an area landmark, they are herein included in the Mount Trumbull Mining District.

Pine Springs Mining District

One of the more well organized mining districts of the Grand Canyon area, the Pine Springs Mining District (named in 1893) was founded with the location of the Ridenour Mine in 1880 by William Ridenour and Charles Spencer. Other prospects were located by other prospectors in tributary canyons and on the Coconino Plateau.

On February 13, 1883, Ridenour and others formed the Colorado Grand Canyon Mining and Improvement Company, recorded by Margaret Ridenour. Ten years later, on May 12, 1893, a meeting was called by nine miners of this general area to form and name the Pine Springs Mining District. Elmer Kane was elected chairman, and L.H. Jewell secretary. Others present were J.E. Davis, W.K. Danson, J.L. Nelson, W.A. Frazier, Alex Hattan, J.H. Farlee, and J.D. Selken.

Sometime after the meeting, prospectors John Conners and Franklin French of Holbrook, Arizona, named the previously untitled tributary Prospect Canyon in order to identify the location of the claims (Coconino County Mining Records, 1893). The Walapai Mining District of the early 1900s, on the Coconino Plateau, is considered to be a southern extension of the Pine Springs district, and thus is included therein in this publication.

Havasu Canyon Mining District

Cataract Creek, one of the easiest ways into Grand Canyon, was a natural focus for the earliest prospectors. By the 1880s, and perhaps sooner, Havasu Canyon was already an active mining area. There had been no formalized district, but had been only a collection of claims located in Cataract Creek (or Havasu Creek, the lower extremity of the same tributary), which had various other local and now-forgotten names. Most of the mining activity was along the Havasu Creek portion of the drainage, near the beautiful waterfalls downstream from the village of Supai, Arizona. Havasu Canyon Mining District, is introduced here as a term to simplify the delineation of boundaries, and to collectively include all claims and mines within the area of Cataract/Havasu Creek. The Havasupai Indian Reservation lies within this district.

Hacks Canyon Mining District

The copper deposits in Hacks Canyon (Hack Canyon on older maps) became a center of mining activity in the early 1900s. A mining district was never established, but the Hacks Canyon Mine was well known. In profitability, the mining area had its ups and downs—mostly downs—until the 1970s, when uranium became a valuable commodity. By the 1970s, several mining claims virtually covered the canyon and the surrounding areas of the Kanab Plateau. Hacks Mining District was used at times for identification purposes only. In this publication Hacks Canyon Mining District is used.

Warm Springs Mining District

During the middle and late 1880s, small copper claims and mines in Warm Springs Canyon on the west side of the Kaibab Plateau were becoming well known as potentially large copper-producers. Ryan, a small settlement, now abandoned, at one time also called Coconino City, became a copper smelting center around 1900. Several claims in the area were identified either as part of the Warm Springs Mining District or the Jacob Lake Mining District. The two districts are very close to one another, and the boundaries often became confused or overlapped. Copper ore was produced from the same rock strata in both areas. Since most of the claims in the area identified the mining district as that of Warm Springs, the name is retained in this publication. It includes all the mining claims that were also in the former Jacob Lake Mining District.

Grand Canyon Mining District

The Grand Canyon Mining District originally encompassed most of the eastern Grand Canyon area. Several copper and asbestos mines were located in this district during the 1880s and 1890s, and prospectors frequently commingled in the area. No one knew what mining district they were in at the time, but the name Grand Canyon was well known.

On February 20, 1893, miners of the eastern Grand Canyon area held a meeting in order to adopt rules and regulations for the location of mines in what was to be known as the Grand Canyon Mining District. W.H. Ashurst was temporary chairman at the meeting held at Cottonwood Canyon. Peter D. "Pete" Berry was elected temporary secretary. A motion, unanimously passed, declared the organization to be permanent.

The original district was formally defined by metes and bounds (Recorder's Office, Flagstaff, Arizona, Record of Mines, Book 1, pp. 225-227): Commencing three miles west of the Bright Angel Trail on the rim of the Canyon; following Horn Creek thence running north to the opposite rim (North Rim); thence running east along the north rim to a point east and opposite the mouth of Red Canyon (Cape Royal) four miles east of the Hance Trail; thence running south to the mouth of Red Canyon, up the natural course of Red Canyon to the South rim, thence west along the rim to the beginning point.

The original Grand Canyon Mining District did not include several mines and prospects west of Horn Creek, those to the east in the areas of Chuar Creek and Nankoweap Canyon, or the asbestos and copper mines at Bass Canyon farther west. For the purposes of this

publication, these mines and claims are included in an expanded Grand Canyon Mining District because they all are in the eastern Grand Canyon region. Thus the size of this district is made comparable in area to other mining districts farther west.

Francis Mining District

In the early 1900s, the Anita Mine and many other associated claims south of the eastern part of Grand Canyon became the center of the Francis Mining District. This district had no specified boundaries.

Little Colorado River Mining District

Perhaps the least active and least known mining district of the Grand Canyon region is the Little Colorado River Mining District. The few placer and uranium claims found in and near the canyon of the Little Colorado River did not specify a mining district name, but mentioned the Little Colorado River. However, Mike Anderson (personal communication to Billingsley, 1995) has a letter from George Tanner which states that Seth B. Tanner created the "Little Colorado River Mining District." Because the Little Colorado River is a large tributary, the claims and mines along it are worthy of a district name of their own. That name is used here. Its eastern boundary is approximately defined along the 111°30' meridian.

chapter two

THE EARLIEST MINES
& CHANGING CULTURES

The Grand Canyon region has a rich cultural history. American Indians, whose livelihood depended upon the land in every way, were obviously the first prospectors in the region. Archaeological information and oral histories tell of the uses of minerals in everyday life, special occasions, and as items of trade between tribes in the Southwest.

In the sixteenth century, the first Spanish incursions brought a strange new perspective on economy, one based solely on the possession of metals and land. Natives and colonists alike saw their respective cultures change forever.

American Indians in the Grand Canyon Region

Before prospectors and early explorers came to Grand Canyon in the nineteenth century, American Indians had lived and roamed in virtually every part of every tributary canyon of the Grand Canyon region. More than a hundred American Indian routes into the Canyon are known, and archaeological remains are abundant both within and around Grand Canyon. The people who lived in the Canyon were highly mobile in the sense that their main concerns were to find food and water, live in a place until these commodities ran low, then move on elsewhere in the region. Even if water was plentiful in earlier centuries, food probably was not. Though people spent more time finding food than minerals, colorful copper minerals and salt were of some interest to them. The following introduction to the American Indians of Grand Canyon has been prepared by Robert C. Euler, former anthropologist and archaeologist for Grand Canyon National Park (personal communications, 1990, 1995, with additional information from Helen Fairley).

When Spanish explorers visited Grand Canyon in the summer of 1540, the first Europeans to do so, they found the land inhospitable and wrote that there was no way for humans to descend to its depths. Until recently, the earliest evidence of human occupation archaeologists had were from Archaic hunters and foragers who, four thousand years ago, lived in rockshelters and made small twig figurines of the animals they hunted. These effigies were left in caves, probably as some form of imitative magic that would ensure success in the hunt. Others painted what probably were religious motifs on the walls of an isolated cave some time between 2000 B.C. and A.D. 1. A single Folsom point was recently found, however, that extends our evidence of occupation of the region perhaps to 10,500 years ago.

A later culture, the prehistoric Pueblo, under favorable climatic conditions, spread throughout the inner canyon and on both North and South Rims by A.D. 1000. They farmed corn, beans, and squash in well watered side canyons, and hunted and gathered a wide variety of animal and plant foods. They knew of almost one hundred routes of access into the Canyon and were on intimate terms with its natural resources. Archaeologists have recorded between 2,500 and 3,000 prehistoric Pueblo ruins throughout Grand Canyon. Some time around A.D. 1200, however, the prehistoric Pueblo abandoned Grand Canyon. The reasons are not known, but an extreme and long lasting drought occurred on the plateaus. The prehistoric Pueblo then may have moved east to the present Hopi country of northern Arizona and the Rio Grande River.

Another group of prehistoric people, known to archaeologists as the Cohonina, occupied much of the South Rim west of Bass Point and on the adjacent sandstone bench within the Canyon we call the Esplanade from about A.D. 700 until 1150. They were farmers, hunters, and foragers whose life style was similar to that of the prehistoric Pueblo, especially in architecture and pottery making. It is not known where these people came from or where they went when they left the Canyon.

Around A.D. 1300 the southern Paiute moved to the north side of the Canyon while the ancestors of the Hualapai and Havasupai (the Pai) spread out over the south side. The latter two still live at the Canyon, the Havasupai in its depths and the Hualapai along the South Rim in the west. The Paiute, forced onto reservations in the late nineteenth century, still regard Grand Canyon as part of their territory, but their present homes are farther north, essentially along the Utah-Arizona border.

During the recording of thousands of prehistoric and historic ruins and the excavation of a few in Grand Canyon, archaeologists have found almost no evidence of aboriginal mining activities. Clay deposits, including kaolin, were sought for pottery making. Salt was mined from two sources: one near the mouth of the Little Colorado River, which the Hopi continued to use until early in the twentieth century, and the other from a deposit now under the waters of Lake Mead. Pieces of malachite and red and yellow ocher also have been found in ruins, and at one site in Nankoweap Canyon a few marcasite crystals were recorded. Calcite and quartz crystals are rarely found, but beads and pendants were fashioned from travertine (including aragonite and calcite deposited as travertine) and fossil crinoid stems. These really cannot be categorized as having been mined, however. The same may be said for turquoise, from which ornaments were made, and obsidian, used in the manufacture of arrow points. Neither of these are found in Grand Canyon but were obtained through trade. Chert, also used for stone tools, on the other hand, occurs in quantity, especially in limestone formations.

Salt Mines

According to legend, in the depths of the Little Colorado River abide the spirits of the Hopi. In the dim past, these legends tell, the spirits emerged from the Canyon, and the dead returned to reside in hadean gloom. This exit from the world beneath is known to the Hopi as the Sipapu. Other legends tell of ghostly inhabitants who, arising from the abyss with glowing eyes and monstrous forms, traveled out across the Painted Desert to revisit their earthly homes on the Hopi mesas, where they now maintain a lively and beneficent interest in human affairs. It is natural, therefore, that these great canyons, the abode of the dead, should be regarded with wonder and reverence, and that

Figure 5. Otis (Dock) Marston, a river runner and Colorado River historian, at the Hopi salt mine near Mile 63. Since this photo was taken, these mines have been closed to the public. Photo by George Billingsley, 1973.

the things found in the canyons are possessed with great mystical powers.

Near the confluence of the Little Colorado River in the Tapeats Sandstone (figure 5) salt leaches out of the sandstone as water percolates through the coarse-grained sands and then evaporates. An ancient trail from the Hopi mesas to the salt deposits travels westward across the Painted Desert to the Little Colorado River gorge, then plunges down to the river through what today is called Salt Trail Canyon, and proceeds downstream. The salt deposits can also be reached by following the Tanner and Beamer Trails. Since prehistoric times, people have made this precarious journey, close to the mystical underworld, returning with their heavy burdens. A man who returned with the salt was considered very brave (Titiev, 1937). Hopi continue to use salt in ceremonies, but changing cultural beliefs among younger generations have diminished the number of collecting expeditions. Today, the National Park Service prohibits visitation to the salt mines. Only sanctioned visits by the Hopi are allowed. The information provided here about the Hopi salt deposits downstream from the Little Colorado confluence was obtained from published sources and from visits made prior to the implementation of more careful cultural protections in 1989.

The Salt Trail is a rough, long journey. The mines themselves are at the base of the Tapeats Sandstone cliff at the greatest angular unconformable contact with the Middle Proterozoic Dox Formation (figs. 1, 5). The Dox is a soft sandstone and siltstone that is easily weathered, and it undermines the more resistant Tapeats, thus creating small (five to seven feet square) protective overhangs in which the salt accumulates. Salt-laden water seeps from the Tapeats, quickly evaporates, and leaves a thickening encrustation of salt on the sandstone cliff and in the small dugout overhangs. Every year many hollow tube-like salt stalactites, about the size of a drinking straw, grow from a few inches to several feet long. Wind accompanying spring cold fronts and local summer thunderstorms breaks these delicate crystal salt drips, which pile up as part of the salt encrustation at the base of the cliff. On a Colorado River trip in 1972, Billingsley noted several noncontemporary red drawings in the area, but the source of the red pigment is probably not local.

George Wharton James, in his delightful 1901 narrative of life in Grand Canyon country, worried even then over the deterioration of the Salt Trail from lack of use. He did not allude to the spiritual importance of the salt deposits, and instead invited a comparison of "their precious salt ledge" with "the superior quality of the article purchased of the Indian traders" (James, 1901:239). James, in pointing out the supposed discontinued use of the Hopi salt, stated that "the trail is rapidly becoming impassable, and unless something is speedily done to it, not even the agile Hopi and their fearless ponies will be able to use it." He is unclear in his narrative as to whether he visited the salt deposits.

During one of Billingsley's expeditions down the Salt Trail in 1968, it was evident that pack animals could not negotiate the trail. During this trek, salt deposits were also found in a northern tributary drainage downstream from the junction of Salt Trail Canyon in the Little Colorado River gorge. This also may have been a source used by ancient people.

On a scientific investigation of springs along the Little Colorado in 1967, Billingsley and party saw three poles of local driftwood, about twenty-five feet long, propped up against the Tapeats Sandstone cliff just upstream from the salt mines. These poles may have aided in climbing down the cliff from a trail above, either in prehistoric times or in the late 1800s when prospectors began roaming this area. The trail above the Tapeats cliff is now referred to as the Beamer Trail, named for a prospector-farmer who lived at the Little Colorado-Colorado confluence around 1890.

In June 1953, Walter W. Taylor tried to locate the Hopi salt mine. Taylor (1954) reported a chemical analysis of salt from near Kwagunt Canyon, determining that one-quarter of it is in fact sodium chloride, common table salt. The rest of the soluble material is a mixture of sulfates and chlorides: 5 percent calcium sulfate, 5 percent magnesium sulfate, 3 percent sodium sulfate, 2 percent potassium chloride, and traces of potassium sulfate and magnesium chloride. The insoluble part of the deposit is made up largely of calcite (travertine) and plaster (gypsum). Other regional salt deposits analyzed by Eiseman (1959) yielded results including 64.2 percent sodium chloride.

The Havasupai also had sources of salt, as indicated in the synthesis of Alfred F. Whiting's studies of these people (Weber and Seaman, 1985). Deposits in Cataract Canyon were red colored and used only in cooking. Purer, white salt from deposits elsewhere in Grand Canyon was used as a condiment.

The supply of relatively pure salt . . . could be reached only with difficulty. One left the rim and descended by a foot trail to a point near the Colorado River. From here a man was let down over a low cliff by means of a rope made of braided strands of bighorn sheep hide. Here the salt hung "like icicles" from the roof of a big cave. Pieces of this rock-like deposit were broken off and placed in sacks in amounts weighing from twenty-five to thirty pounds. . . . Along the Little Colorado River there were several salt deposits; the river was, therefore, called . . . "salty river." (Weber and Seaman, 1985:159-160)

Salt can be found at the base of the Tapeats Sandstone cliff in many places in Grand Canyon, wherever water seeps out of the sandstone. Such seeps taste salty, and in protected niches they develop delicate, hollow, hanging salt tubes. In places where seeps are active all year, salt stalagmites as well as stalactites will form. Salt can also be seen leaching directly from rocks underlying the Tapeats, including the Vishnu Schist. In the Vishnu, Tapeats salt can pass through joints (as seen by Spamer in 1991 along the Colorado River at Mile 77.4 left) or even through the foliation of the schist (as seen along 128 Mile Creek in 1994 by Spamer and Richard Quartaroli).

Other sources of salt were elsewhere in the greater Grand Canyon region, particularly in the valley of the Virgin River, Clark County, Nevada (now drowned by the Virgin Arm of Lake Mead), including the Salt Caves near the former site of St. Thomas, and outcrops of salt over a distance of more than twelve miles between St. Thomas and the former confluence of the Virgin and Colorado Rivers (Phalen, 1919; Harrington, 1925, 1926). An especially notable source was Salt Mountain, or Big Salt Cliff (now called Salt Point), three miles south of St. Thomas. An American Indian salt mine there was first described by fur-trapper Jedediah Smith in 1826. The exposure had been a half mile long and 100 to 150 feet high (Carlson, 1974). The salt had been used in especially large quantities as a flux in ore processing at the mills at Silver Reef, but in a good example of the principle of supply and demand, freighters hauling the salt received just twenty-five dollars per ton, despite the 235-mile round trip over bad roads (Larson, 1979:174-175).

Red Pigment Deposits

The Pine Springs Band of the Northeastern Pai of western Grand Canyon found a cave that contains a rich deposit of red mineral pigment, mostly hematite. This pigment, when mixed with deer tallow, makes a face cream and body coating that protects the skin against sunburn or cold. Its skin-protecting properties, and its bright red color, made the pigment a precious commodity that the Pai traded for other valuable goods.

Until the late 1820s, Pai traders exchanged the pigment with the Panya people to the west for shells which the Panya had received from Pacific coast tribes in trade for red pigment. The Pai, in turn, traded their shells for other commodities with tribes to the east. The Hopi, Zuni, Acoma, and other people believed the ocean shells to be important in many weather control ceremonies, and some shells were used for jewelry. *Olivella* shells have been found occasionally in Grand Canyon archaeological sites (e.g., Euler, 1978). About the shell and red pigment trade, Dobyns and Euler (1976:25) have remarked: "They [Pueblo peoples] obtained much of the Pacific coast shell supply from Pai traders, who undoubtedly turned a profit on such transactions. The Pueblo trading partners of Northeastern Pai entrepreneurs paid for shells and red pigment with foodstuffs and craft products. . . . Thus, the Northeastern Pai people produced a significant economic surplus to trade east and west to other tribes."

Later, Spanish travelers used the red pigment, having obtained it from Pueblo suppliers. By 1827 to 1830, merchants ranged west from the Rio Grande country into Northeastern Pai country freighting cheap United States manufactured goods, ending the non-native demand for the red pigment (Dobyns and Euler, 1976), although in 1941 a generously measured teaspoonful of pigment would sell among the Pueblos for twenty-five cents, or larger amounts for cash or goods (Weber and Seaman, 1985:162-163).

One of the best-known sources of the red pigment is a shallow overhang cave in the Redwall Limestone, just large enough for one person to enter (Weber and Seaman, 1985). Mining the red pigment was a dangerous, difficult process. Tall ladders or notched logs were constructed and propped against the nearly vertical canyon wall, in order to climb from the nearest ledge into the cave. The cave was formed by a dissolution of the limestone by ground water, far below the ground surface, about 310 million years ago (Billingsley, 1986). About 300 million years ago, during the time that the Surprise Canyon Formation was being deposited atop the Redwall, this cave, as with others that had developed in the Redwall, was filled with iron-rich, sediment-laden waters. The fine-grained Surprise Canyon silts accumulated in thick layers which oxidized over a long period of geologic time. The oxidation process converted the greenish iron-rich siltstone to a bright red color wherever sufficient iron minerals were present. When natural erosion of side canyons cut through the Redwall Limestone, the silt-filled cave was exposed for the American Indians to find. Today, the cave forms a bright "red eye" about 450 feet down the Redwall cliff, where it remains a sacred area to the Hualapai and Havasupai people.

As summarized by Weber and Seaman (1985), other sources of red pigment have been said to be along Diamond Creek and in Meriwitica Canyon (Kroeber, 1935), east of Cataract Canyon (Weber and Seaman, 1985), and unspecified deposits on the north side of the Colorado River (Dellenbaugh, 1934).

Other American Indian Mineral Uses

A.F. Whiting's noted ethnographic studies of the Havasupai, who traditionally have called the inner canyon home, includes useful observations of the incorporation of minerals, metals, and rocks into their culture (Weber and Seaman, 1985). Chert was used with a flint knife to produce a spark for fire. Pottery clay was found near Pine Springs, in the area occupied by the Hualapai Tribe, and a deposit of hard clay was mined near Hermits Rest on the South Rim west of Grand Canyon Village. Similarly, sphalerite and galena (zinc and lead minerals), used to make black paint, came from near Pine Springs. Copper minerals from near Grandview Point were used for ornamentation and as trade items with the Hopi. Flint and obsidian were used to fashion knives and arrowpoints. Gypsum was mined near Supai and near the head of Lee Canyon. After processing with heat and water, it was used as a white body paint. The Havasupai knew of iron pyrites, or "fool's gold." Weber and Seaman (1985:162) cited no uses for the pyrites, but they did mention that it was known to occur at the Bass Mine. An unknown mineral was used to make blue paint. "This mineral, no longer used in 1941 [when Whiting was studying the Havasupai], came from a deposit in Grand Canyon. Formerly it was used in face painting by both men and women, particularly for dances" (Weber and Seaman, 1985:164).

The Europeans Arrive

Non-native people first visited the Canyon (briefly) in September 1540, when Garcia Lopez de Cárdenas, a few other Spanish soldiers, and several Hopi guides reached the South Rim probably near what today is Desert View (Winship, 1896; Hughes, 1978). They were an excursionary party sent from the main force at Zuni Pueblo (New Mexico) under the command of Francisco Vásquez de Coronado, after Pedro de Tovar had returned there with Hopi stories of a great river to the west of the Hopi pueblos. The expedition had been in the area to locate the so-called Seven Cities of Cíbola, ostensibly rich in gold, which were to be added to the Spaniards' systematically acquired plunder. After a twenty-day journey to the west, Cárdenas and party by all accounts were discouraged with Grand Canyon. They left after failing to find a way down to the Colorado River; a notable, and perhaps intentional, oversight on the part of the Hopi guides. Non-native visitations remained sporadic at best well into the nineteenth century.

James (1901:67) makes a singular, almost casual remark without considering its historical implications: "I am informed that it was recently discovered, in looking over some titles in the old registrar at Tucson, that silver mining was extensively undertaken near Red Butte [a prominent landmark on the Coconino Plateau, about twelve miles south of Grand Canyon Village] by the Spaniards about the year 1650 and later." Of course, the Spaniard's quests for riches are historically well known and form an integral part of most post-Columbian histories. If James's source is accurate, this could be one of the earliest records of non-native mining in the Grand Canyon region, but there is no corroboration of—or for that matter, source for—his statement. No evidence of mining has been found near there, and the likelihood of any silver deposits at or near the surface there is very remote, even though the Mesozoic strata that compose

the butte are the same age as silver-bearing strata in Utah, north of Grand Canyon. Today, silver and uranium ores are found in a buried breccia pipe (Canyon Pipe on Map 2) about four miles northwest of Red Butte, but the silver is not present in economical concentrations, and in fact it was not known at all until uranium exploration companies discovered the ore deposit by exploratory drilling methods in the early 1980s.

In 1858, the first real exploring expedition reached the western part of Grand Canyon. Under the command of Lieutenant Joseph C. Ives, Army Corps of Topographical Engineers, the expedition was to ascertain the resources of this territory. Ascending the lower Colorado River by steamer from Yuma, Arizona, the expedition on March 12, 1858, reached the head of navigation in Black Canyon (downstream from present-day Hoover Dam). The expedition then trekked overland, reaching Grand Canyon near Diamond Creek on April 3. Although Ives, like his Spanish predecessors three centuries earlier, was dismayed by the Canyon, the expedition's physician, John Strong Newberry, trained also in geology, made the first comprehensive geological studies of Grand Canyon. Nowhere in Newberry's (1861) pioneering geological report does he mention any economically significant mineral occurrences in the Grand Canyon region.

About this time, the first itinerant travelers probably began to wander into Grand Canyon. One could infer that European-Americans were indeed casually wandering into the Canyon by the time Ives and his party arrived there in 1858. Ives (1861, Part 1, p. 100) noted in his journal on April 3, the day they reached the Canyon:

Our party being, in all probability, the first company of whites that had ever been seen by them [the local people], we had anticipated producing a great effect and were a little chagrined when the old woman, and two or three others of both sexes that were met, went by without taking the slightest notice of us. If pack-trains had been in the habit of passing twenty times a day they could not have manifested a more complete indifference.

But this is not to say that Grand Canyon transients were prospectors, even though few others would have had the reason to wander into the canyons. Unfortunately, if prospectors were there at the time, there are no known claims from that early date.

Any early prospecting activities in the Canyon were hardly likely to have brought much attention from the outside world. In hindsight, we know that no marvelously rich,

Figure 6. The Diamond Creek Hotel (or Farlee Hotel) as it appeared around 1915. No longer standing, it was at the confluence of Peach Springs Canyon and Diamond Creek, about one mile from the Colorado River. Photo courtesy of Mohave County Historical Society, Kingman, Arizona, photo no. 682.

long-lived deposits of precious and base metals are present in Grand Canyon, and there were no nearby settlements that could handle bulk processing of newly found ores. It is possible that the itinerant presence of prospectors might fit well into the overall picture of native "indifference"; however, the miners were soon to contribute to the grief of the native peoples of Grand Canyon, through displacement and cultural erosion.

The First Hotel

By the 1880s, Julius H. Farlee and a partner known only as Young, made up their minds that tourism was more profitable than any mineral they ever found, and they built the first hotel at the Canyon, certainly the first commercial residence inside the chasm. This mining-to-tourism switch was to be repeated by several early Grand Canyon prospectors.

The first tourist route into Grand Canyon was the same one followed by Lieutenant Ives and party in 1858, twenty miles down Peach Springs Canyon to its confluence with Diamond Creek. From there it was a short, easily traveled distance to the Colorado River. The Diamond Creek Hotel, or the Farlee Hotel (figure 6), operated from 1884 until perhaps 1901.

MINES & PROSPECTORS
OF THE GRAND CANYON REGION

Recorded mining claims in the Grand Canyon region date from the late 1860s through the middle of the twentieth century. The pattern of prospecting can be determined from county records of Mohave, Coconino, and Yavapai Counties, Arizona. Some North Rim locations are recorded in Washington County, Utah. Unfortunately, the movements and activities of many early prospectors were never recorded. Their claims were made by word of mouth or by pieces of paper left in cans. Passage of the Federal Mining Law in 1872 specified formal procedures for locating and recording claims, but many prospectors seem to have just collected what ore they found from time to time, never filing a claim. The mining law, however, did instigate better records for a few of the more major mining activities in the Grand Canyon region. But far less is known of the history of mining in this region than in the mineralogically richer, more productive, politically more incendiary, and ofttimes socially rarefied frontier mining communities of central and southern Arizona. In fact, the two volumes of the History of Mining in Arizona (Canty and Greeley, 1987, 1991) and Spence's (1993) history of mining engineering in the American West, completely ignore the Grand Canyon region, but they do provide a historical frame of reference for the meager operations around Grand Canyon. Arizona Odyssey, Goodman's (1969) outstanding categorized bibliography of nineteenth century magazine articles about Arizona, lists an incredible number of articles about mining in southern and central Arizona already in print well before the first of the relatively few about northern Arizona. Garbani's (1993) listing of Arizona Mines and Mining Companies includes only a few of the mines and mining areas of the Grand Canyon region. And from the perspective of world-class mineral deposits (prior to the uranium booms) Hausbrand et al. (1927) published a series of eight maps of the world, depicting the occurrence of major mineral deposits. In the Grand Canyon region (on sheet 5), a single symbol in the eastern Grand Canyon depicts *asbest* (asbestos) and its mode of occurrence,

Kontaktlagerstäten (contact-metamorphism deposit). Presumably, the symbol maps both the Hance and Bass asbestos mines, although at the map scale a symbol for each could have been accurately placed.

The Grand Canyon mines were virtually unknown or considered to be inconsequential to the mining community prior to the discovery of uranium, except when their involvement in political and judicial conflicts brought them into the general public's eye. Such anonymity also fell upon the men and women whose lives were touched by Grand Canyon's meager mineral riches.

In this chapter we provide historical and cultural information on the mines and prospectors of each of the mining districts of the Grand Canyon region. In a few cases, all that is known can be written in a sentence, but we are aware that these historical tidbits are, nonetheless, parts of the greater story of the Canyon's human history. In time, researchers may serendipitously find new information and revise the things we present here. These will be welcome additions to the history of this spectacular and unforgiving region.

Hazards of Early Canyon Prospecting

Once prospectors reached the rim of Grand Canyon, they were faced with several immediate questions: Now what? How do I get down? Are there precious metals? Is there any water? It was the last question that was the most desperately in need of an answer.

The climatic extremes reached in the Canyon and on the adjacent plateaus are dramatic and life threatening. During winter months, temperatures drop below zero degrees Fahrenheit in the rim country, and occasionally drop even that low along the Colorado River. The heat of summer in the Canyon depths can reach 125 degrees; rocks become too hot to touch for more than a few seconds. This tends to discourage prospecting. So, the Grand Canyon prospector's schedule probably divided time between the plateaus during the warmer and temperate seasons, and the inner canyon during the colder months (Kolb, 1914, has mentioned this mode of seasonal migration of prospectors).

Within the depths of the Canyon, the climate is that of a desert. The average yearly rainfall is less than ten inches. The surrounding higher plateau country, six thousand to nine thousand feet in elevation, has an average annual rainfall of fifteen to twenty inches. On the plateaus there are

Early prospectors who were not so discouraged by the Canyon's immensity that they moved on to more familiar terrains must have found the logistical problems of getting around in the Canyon considerable. Burros were commonly used as pack animals because horses, and sometimes mules, could not take the continuous strain of travel inside the rugged canyon. Those bringing along pack animals had a

Figure 7. "Flagstaff, A.T. Gold Miners Packed For the Grand Canyon." Photo courtesy of the Museum of Northern Arizona, Flagstaff.

very few perennial water supplies for nearly a hundred miles in most directions from the canyon rim. What rain and snow that does fall soon passes directly into the porous limestones that form the plateau surfaces. Just to get to the rim, prospectors may have had to carry several days' worth of food and water for themselves and their pack animals (figure 7). In many cases, we will see that water was a far more valuable commodity than any mineral riches the prospectors and miners may have dreamed of. The Colorado River, of course, is a plentiful and reliable source of water, but it can be safely reached only through a few widely separated tributary canyons. Many discouraged prospectors gave up their pursuit of riches and concentrated on getting back to the nearest settlement alive.

doubly hard time of getting around. Just a few major side canyons provided easy routes into the Canyon: the broad Peach Springs Canyon (down which today a seasonally improved vehicle road descends to the Colorado River) and Cataract/Havasu and Kanab Canyons which are reasonably passable routes into the inner canyon.

Natural foods in this region are hard to find and are available only in limited amounts. A person cannot linger long in one area. A prospector could even starve to death if too much energy was expended in the search for food. Thus many of the abandoned miner's caches throughout Grand

Canyon contain the relics of packed-in foods and the implements necessary to prepare them.

The prospector's pack animals needed food, too, an additional burden in the long marches through the Canyon. Some prospectors abandoning the Canyon also abandoned (or, in death, left behind) their burros, which developed feral populations. The burros' progeny became an ecological nuisance in the latter part of the twentieth century, competing for the same resources used by bighorn sheep and other indigenous animals. The feral burros have been captured and removed from the national park, mostly during extraction programs conducted by the National Park Service and private animal humane societies in the 1970s. Today burros are still found in several remote canyons of the Hualapai and Havasupai Indian Reservations.

Popular images, created by authors and movie producers, of grizzled, lone prospectors with burro, pick, pan, and coffee pot were hardly artificial. The rugged terrain of Grand Canyon called for just that sort of prospecting operation and personality. Location of a minable prospect was followed by tedious, backbreaking work, usually by one man, sometimes two or three. Many of the operations in Grand Canyon were subsistence mining, and only a few commercially viable mines saw systematic development. Even the twentieth century's modern amenities were available only to the more easily accessed plateau-based mines.

Modern techniques were impossible in the small, inner-canyon operations. Drilling, cutting, hoisting, and pumping were accomplished with hand tools only; in some, perhaps also with the labor of burros. Since electricity was unavailable, lanterns were the norm. Ventilation of the mine tunnels and shafts was solely by natural air currents, so the air was probably bad most of the time. Electric explosives were impractical, too. When explosives were used at all, light-and-run methods were employed.

Timbering was almost unknown in the inner canyon mines because there was no practical way to transport enough wood to do the job. Fortunately, most of the mines were driven into fairly solid rock, and while the mines were still new, men were not placed in as much jeopardy as one might suppose. However, one of the prospects near the mouth of Lava Canyon in the Palisades Fault is partly shored up with driftwood timbers (see under Little Colorado River Mining District), a supply of which was once plentiful along the Colorado River.

The mines were claustrophobic, lonely, and dank. Not until larger, more professional operations were started in the plateau mines were more efficient and safe means of work made available. The fate that came to some men was recounted by Ellsworth Kolb (1914:224-225), who provided an enduring but lonely epitaph for one anonymous Canyon pioneer prospector:

We camped a short distance below the rapid [downstream from Pipe Creek], just opposite a grave of a man whose skeleton had been found halfway up the granite, five years before. Judging by his clothes and hob-nailed shoes he was a prospector. He was lying in a natural position, with his head resting on a rock. An overcoat was buttoned tightly around him. No large bones were broken, but he might have had a fall and injured himself internally. More likely he became sick and died. The small bones of the hands and feet had been taken away by field-mice, and no doubt the turkey-buzzards had stripped the flesh. His pockets contained Los Angeles newspapers of 1900; he was found in 1906. The pockets also contained a pipe and a pocket-knife, but nothing by which he could be identified. The coroner's jury—of which my brother [Emery] was a member—buried him where he was found, covering the body with rocks, for there was no earth.

Such finds are not unusual in this rugged country. These prospectors seldom say where they are going, no track is kept of their movements, and unless something about their clothes tells us who they are, their identity is seldom established. The proximity of this grave made us wonder how many more such unburied bodies there were along this river.

Bentley Mining District

Before the Ives expedition in 1858, prospectors probed along the extreme western edge of Grand Canyon north of the Colorado River. Gold was sought, but it proved elusive except in a few placer claims farther west from the Grand Wash Cliffs. It was just a matter of time, though, before the meager mineral resources of Grand Canyon country were exploited.

Grand Gulch Mine

Although Hill (1915) has said that copper deposits in the Grand Gulch area were discovered around 1853, by Richard Bentley, Samuel S. Adams, and others, the information is not correct. Altschul and Fairley (1989:209) reported that W.P. Jennings, "a former manager and part owner of the mine in the early 1900s," recalled that around 1871 the copper had been discovered by a Shivwits Indian who told some residents of St. George, Utah. A claim for

Figure 8. The Grand Gulch Mine around 1904, showing smelter stack, headframe, and tailings. The smelter stack masonry is composed of mud, bricks, and local limestone. The view looks northwest on the Sanup Plateau. Photo courtesy of the State of Arizona, Department of Library, Archives and Public Records, Phoenix, Arizona.

the Adams Lode in the Bentley Mining District was filed in Phoenix by Richard Bentley, Samuel S. Adams, Erastus Snow, James Pearce, and eight other men.

The copper deposit was on the Grand Gulch Bench (Sanup Plateau) near Pigeon Canyon, about twelve miles north of the mouth of Grand Canyon. The Walter Jennings family of Salt Lake City, Utah, handled the financial aspects of the business. In 1873, H.C. Kiesel, a Salt Lake City clerk, had traveled down Grand Wash to the Colorado River. He mentioned in correspondence sent to the Salt Lake Tribune (Kiesel, 1873, as typeset):

The mine of Messers Adams, Snow and Bently, lays about eight miles each [east] of Pah Koon warm water springs). Professor Thompson went over with them and saw the mine. His opinion is that it seems to be a pocket of mineral in sand stone, and probably would peter out as soon as they go down on it twenty-five or thirty feet. They have worked on it some, just enough to hold it.

According to Othel Milne, who worked at the Grand Gulch Mine (figure 8) for several years beginning in 1902,

the prospect was sold to Adams and Cunningham by an American Indian who had some copper-rich rocks from the Grand Gulch Bench area west of Pigeon Spring (at the head of Pigeon Canyon). In the early days, access to this property was by horse. By wagon, it took several days to travel south to the mine from St. George, Utah, about ninety miles; six dollars was the one-way fare (Cox, 1982:309).

The Bentley Mine, originally recorded in Washington County, Utah (at St. George), was reassigned to Mohave County, Arizona (recorded in Kingman), on June 23, 1873, forming the Bentley Mining District. New descriptions were recorded in Mohave County by Richard Bentley, under the names of Samuel Cunningham, Sam O. Crosby, James Pearce, W.E. Dodge, Erastus Snow, Richard Bentley, Samuel S. Adams, William H. Coranch, Joseph Birch, and B.H. Paddock. An adobe smelter (figure 8) was built to process ore from the Bentley Mine, but it never worked successfully.

In 1882, county tax records indicated, "Grand Gulch Mining Company, 21 acres of patented land, known as Adams Mine was sold for delinquent taxes May 16, 1881, and was redeemed by C.C. Bradley on February 3, 1882, for $6.97." Frank Jennings was the agent for the mine in 1886, and "owners unknown" was listed for 1890, 1891, and 1892. Tax records for 1893 show "20 patented acres, Jennings Brothers, mining claim known as the Adams claim" (Malach, 1975).

The "Adams" prospect was worked intermittently for sixteen years. Thomas W. Jennings later named the prospect The Grand Gulch Mine. From 1906 to 1958, there was only occasional production of copper ore. In 1911 it was known as the richest copper mine in Arizona; between 1906 and 1917, copper concentrate of not less than 14 percent copper was shipped. The remote location of the mine lent to the relatively high percentage of copper; the ore had to be well sorted to reduce haulage costs and effort. After 1910, some, if not most, of the ore was hauled forty miles to St. Thomas, Nevada, a site now beneath the Virgin River arm of Lake Mead.

By 1913, a shaft had been sunk five hundred feet below the surface, into the center of the generally circular ore body with a pear-shaped profile (a breccia pipe, but not recognized as such by the early miners). The ore body is vertically limited to a depth of less than 250 feet (Hill, 1915) in the Pakoon Limestone. It has produced brochantite, azurite, chalcocite, chrysocolla, cotunnite, copper-descloizite, malachite, and limonite. An article in the Mohave County Miner (July 25, 1914) included a plan and profile of the mine, as well as a detailed description of the operation and mineralization. It reported that W.T. Schaller of the U.S. Geological Survey had identified azurite, brochantite, chalcocite, chrysocolla, cotunnite, cuprodescloizite, malachite, and limonite.

When early activity at the mine increased in the 1870s, Harrison Pearce helped establish Pearce Ferry in 1876 (the spelling was later corrupted to Pierce), the Colorado crossing that had been established in 1862 by Jacob Hamblin (Granger, 1983). In 1887, Pearce also operated a dairy at the little sawmill town of Parashant, just north of Mount Dellenbaugh on the Shivwits Plateau.

In the early 1900s, the buildings at the Grand Gulch Mine consisted of a three-room bunkhouse made of heavy rock and a house with three main rooms—one for the office, one for dining, and one for women's quarters. The main engine at the mine was an old-fashioned one-cylinder, four-cycle Fairbanks-Morse (Cox, 1982).

Pat "Vivian," Abraham, and James Bundy hauled freight for the Grand Gulch Mine in 1917. Pat (personal communication, 1974) recalled that the wagon carried 1,200 pounds of ore each trip, and that a mine shaft operator, Bert Snow, accidentally set fire to the mine shaft in that year or early in 1918. The fire spread to the underground timbers, finally burning out the mine. Most of the ore had been mined out by that time, so activity ended for a while, but resumed sporadically until about 1958. Pat Bundy also

mentioned that seventy-five men and women were employed at the mine in 1917: "You could make a fortune sorting ore there for $45.00 per month, instead of cow-punching for $35.00 per month."

Othel Milne recalled that around 1917 or 1918 about eighty men and women were working at the mine. The women were cooks and waitresses (Cox, 1982:310). "Mrs. Riding, the two Blair girls, Jane Larson, Mary Lamb, and Will and Lamar Pearce's wives are some that I recall. And I did love Maggie Averett's raisin pies—she was a wonderful cook. There was also the lady from Los Angeles and a few from down on the Muddy [River] in Nevada. Some of the women had small children out there with them."

By 1951, production at the mine totaled 15,701 tons of ore containing 24,349 ounces of silver, 6,651 pounds of copper, and 715 pounds of lead (Lane, 1984). LaDell Jessup remembers that "the old Grand Gulch Mine dump was reworked from 1955 to 1961. Sam Mosher, a Signal Oil and Gas Company millionaire, financed the operations. They also drilled and tested around in search of other ore bodies, both there and at the Copper Mountain site [in Parashant Canyon about twenty-eight miles farther west]. Some ore was reclaimed from both places" (Cox, 1982:314).

The Grand Gulch Mine has been derelict since 1958, but the uranium boom of the mid-1970s aroused some new interest in it. U.A. Small and C.H. Englehart of American Fork, Utah, owned the mine in 1978, but they concluded that most of the minable ore had been removed. In 1978, two large late-1940s ore trucks were still at the mine, with rotted flat tires and only a few missing parts). One mine building was in good shape, but the others had fallen down, and most of the mining equipment lay scattered and rusting. The main buildings had burned down in 1955.

During the 1880s, the Sanup Plateau yielded still other ore deposits. Many prospects were located by B.M. Ellenbeck and C.A. Smith.

Savanic Mine

On April 20, 1878, James R. Cunningham located the Bronze L Mine in the Wescogame Formation about three miles south of the Grand Gulch Mine in a small canyon that drains westward through the Grand Wash Cliffs. It was acquired by George H. and C.W. Dodge and A.R. Whitehead on June 13, 1880, who renamed it the Savanic. In 1906, the mine was controlled by Levi Syphus and Harry Gentry (Dunning and Peplow, 1959).

Unlike the Grand Gulch Mine, the Savanic is located on a small fault and joint zone on the edge of a breccia pipe.

The pipe itself is well concealed, but the mine has the potential to be an active producer again. The U.S. Bureau of Mines credits the Savanic with producing three hundred thousand dollars worth of copper and silver between 1906 and 1919 (Elsing and Heineman, 1936:96). No subsequent production records have been found.

In 1959, W.E. Covey of Las Vegas, Nevada, owned the mine. There is no evidence that any work was done during the 1960s, and by 1978 the Savanic was owned by the 5M Corporation of Hurricane, Utah, as a possible uranium prospect. As of April 1983 mining still had not resumed, but the 5M Corporation, together with Uranez USA, Inc., of St. George, Utah, planned a joint uranium exploration near the Savanic and Cunningham Mines (Billingsley et al., 1986).

A mineral resource study of the Savanic indicates copper minerals including azurite, malachite, chalcopyrite, bornite, and chalcocite as disseminated stains, fillings in solution cavities, and fracture fillings forming an ore-mineral matrix between the breccia blocks (Billingsley et al., 1986). The highest copper content sampled yielded 18 percent copper; the highest uranium oxide content was 19 parts per million (Lane, 1984). The copper ores azurite, malachite, chalcopyrite, bornite, and chalcocite occurred only in small amounts. Hematite is commonly associated with the copper minerals, especially in large cavities up to two feet in diameter that are wholly or partially filled with these minerals.

Cunningham Mine

Most likely (but not certainly) the Cunningham Mine was located as only a copper prospect about the same time as the Savanic, a mile distant, by James R. Cunningham in April 1878. The Cunningham is at the head of a northern tributary drainage to Snap Canyon, in the Pakoon Limestone. The first record of this prospect being called the Cunningham Mine was discovered during a 1982 study for potential mineral resources of the area (Billingsley et al., 1986). Documents on the exploration efforts for uranium on this property in 1978, by the 5M Corporation and Uranez USA, there referred to the scattered diggings as the Cunningham Mine.

It is not known how much copper ore was produced from the small shafts and tunnels of this area. We speculate here that what little was produced was combined with ore shipments from the Savanic Mine. A mineral assessment by the U.S. Bureau of Mines in 1984 yielded 4.2 percent copper from the Cunningham ore deposit; the highest uranium oxide content was 48 parts per million, and the highest barium content was 6.1 percent (Lane, 1984).

Hidden Canyon Mines

About fifteen miles north of the Grand Gulch Mine is Hidden Canyon, a large tributary draining the Shivwits Plateau. Mines are located at the northern part of the Sanup Plateau, on a hill in the Hermit Shale near the contact with the overlying Esplanade Sandstone. They are stratigraphically in the same position as the Grand Gulch Mine, but at an elevation of about 4,200 feet. The hill is bleached yellow, unlike the usually red Hermit Shale, revealing the breccia pipe on its north side into which three mine tunnels are found. In some of the tunnels are found the usual copper minerals azurite, malachite, and an unidentified light-blue copper mineral. Associated with them are hematite, limonite, calcite, and minor amounts of chalcocite. Uranium oxide has been measured at 37 parts per million in these ores, and in another sample at 4.5 percent (Lane, 1984).

Very little is known of the history of these mines. They may be instances of miners taking the copper ore without filing claims. An old wagon road passes through the limestone narrows of Hidden Canyon west of the mine, where also several ancient American Indian pictographs can be found.

On April 14, 1982, Pathfinder Mines Corporation made claims on the Hidden Canyon mines and began exploratory drillings for uranium. In 1982, Energy Fuels, Inc., also staked uranium claims near the Hidden Canyon mines, but nothing came of them. Smaller-scale drillings had been made in 1976 by an unidentified party. Uranium ore was found in small pockets by the Pathfinder drillings, but they were not economically viable (Pathfinder Mines, personal communication, 1991). Although roads had been bulldozed on the hill to permit the drilling rigs to be set up, they have since been graded away to prevent further access. The longest, lowest of the mine tunnels remained open in 1991, reaching about 120 feet into the hill.

Fort Garrett

One of the mysteries of the western Grand Canyon area is a small stone cabin at the head of Pearce Canyon, seven to eight miles south of the Grand Gulch Mine on the Sanup Plateau. On the USGS 7.5 minute Snap Canyon East quadrangle the cabin is labeled as a ruin, Fort Garrett. A rock pool in the drainage below the cabin holds water, but otherwise there is nothing nearby which gives a clue as to why the cabin is in this lonely spot. Billingsley and Karen Wenrich, while collecting basalt in the area for age determinations in the fall of 1987, found a mining claim on a basalt dike about two miles southeast of the "fort." In an old tobacco can in a pile of rock, the claim located Fort Garrett 2 in the Grand Gulch Mining District (Bentley Mining District). It was signed and dated August 31, 1935, by W.H. Garrett (figure 9). More information was obtained by Menkes from LeRoy Condie (personal communication, 1988) and Reed Mathis (personal communication, 1989), which has helped establish a sketchy history of Fort Garrett.

William "Bill" Garrett was an itinerant cowboy and sometime prospector. Supposedly, he was born in West Texas and drifted to this part of Arizona from Wyoming. From time to time, Garrett fraternized with Bill Shanley, the notorious outlaw of the Arizona Strip. According to one source (Harris, 1980), Garrett's uncle was Pat Garrett, the man who killed Billy the Kid. Garrett was considered to be an excellent cowboy and horseman, and participated in local rodeos. He helped those who ran cattle on the strip, including one Jack Wiggins. One season Garrett helped Wiggins build a shelter. The walls were put up, but there was not enough wood remaining for the roof, so they just stretched a tarp over it. Later, Wiggins recalled the season on the Sanup Plateau and mentioned that he and Garrett had built a "fort" that he named Fort Garrett.

Bill Garrett and Art Coleman were partners in a claim at Gold Butte, Nevada, where they are buried. It was only a modest claim, from which they made just enough to keep them in beer.

As of December 1989 none of the copper mines or prospects on the Sanup Plateau were operating. The 5M Corporation and Uranez USA concluded that there are not sufficient uranium reserves in these claims to warrant any mining in today's economy.

Music Mountain and Lost Basin Mining Districts

During the period of the great federal explorational surveys of the American West culminating in the 1870s, Lieutenant George Montague Wheeler stands out as a far-ranging intervenor in mapping and mineral surveys in the Southwest (Dawdy, 1993). His work in Nevada and California is especially notable, but surveys under his direction also led to the western Grand Canyon and include a remarkable man-powered upriver exploration of the Colorado from the Grand Wash Cliffs to Diamond Creek.

Wheeler was involved in controversial mining ventures that today would be branded as clear conflicts of interest, and he—and therefore the federal government—was accused of privateering in making mineral claims in the areas he mapped and surveyed south of the Colorado River in Arizona, mostly in the Hualapai Mountains. By 1871, several of the claimed mining districts were mapped into what are defined here (Map 2) as the Music Mountain and Lost Basin Mining Districts. Specifically, Wheeler's districts named Andrew Jackson, Gold Basin, Lost Basin (different from our definition), and Pine Springs (greatly expanded herein). Dawdy (1993) has briefly discussed the activities of Wheeler and his operatives, and has provided sketch maps of the mining districts as known to Wheeler (see also Wheeler, 1871).

Some of the old mining ventures in these two districts overlap, so we examine them together. See Map 2 for more precise data on the locations of mines and prospects.

Music Mountain Mine
West and southwest of the Sanup Plateau and the mouth of Grand Canyon, a few isolated gold mines were located in the Hualapai Wash area by desert prospectors who

Figure 9. Claim left by William H. Garrett, dated August 31, 1935, found by George Billingsley and Karen Wenrich in the fall of 1987 in a Velvet brand tobacco can on a basalt dike in the Hermit Shale southeast of Fort Garrett. The claim located the "Fort Garrett 2" lode, in the Grand Gulch Mining District (Bentley Mining District). Photo by George Billingsley.

supplemented their income by working as cattlemen. Gold and copper prospects were claimed as early as 1886, most of which were owned by Mike Scanlon of Basin, Arizona, and Cy Childers of Kingman, Arizona. They had to haul water to their mines and others, and it is likely that they made more money in that venture than they did digging for gold.

Some copper deposits are scattered from a point near the middle of the region nearly to the summit of the Grand Wash Cliffs. Several of these claims were made by James Burrows and J.W. Mouat of White Hills, Arizona. The ore contained some gold and silver, and assayers measured 17 percent to 20 percent copper (Schrader, 1909), but the placer gold mines formed the nucleus of the Lost Basin Mining District.

Jim Music, for whom the mountains are named, was a prospector who, with Sam Crozier, Bill Ridenour, A.E. Davis, Bud Grounds, and others, developed the Hackberry Mine about 1875 near Hackberry, Arizona, to the south. Originally called the Crozier or Ridenour Mine, the Hackberry was named for the hackberry tree growing at the spring below the original claim (Malach, 1975). The mine contained a "rich vein of silver ore . . . from 10 to 18 inches thick" that produced $340 of silver per ton of ore (Raymond, 1877:352).

In 1879 or 1880, gold was discovered by Dave Southwick, Ed Burk, Bill Hatch, and Joe Prisk along the Grand Wash Cliffs below Music Mountain north of Hackberry. Several mines produced gold, one being the Ellen Jane Mine. The ore is found in quartz veins associated with diabase dikes that intruded Proterozoic granites and schists similar to those of the Lower Granite Gorge of the Colorado River. Today, the collection of mines in the diabase dikes are referred to as the Music Mountain mines (Wenrich et al., 1989b).

Placer gold claims were located in various parts of Grand Canyon, but none panned out. In western Grand Canyon, south of the Colorado River, William B. Ridenour located several placer claims on April 13, 1882, on the eastern slopes of the "Truxton Mountains" (Music Mountains), six to eight miles west of "Spencer's" stock range at Milkweed Canyon. No record exists of any placer gold or other mineral finds there. The Lost Partners Mine, "Approximately 74 miles below Mt. Trumbull on Colo. river," was a placer gold operation owned by Louis Arnold of St. George, Utah (Arizona Department of Mineral Resources, Mine Owners Report, June 9, 1941). His report indicated that the mine workings contained an "Old rocker and 2 sluice boxes left where former workers quit work and quarreled over supplies . . ." Arnold elaborated:

According to old timers there were three men from Pipe Springs working this bar and due to falling out over going for supplies they fought and broke up. No one knew where this claim was and plenty hunted but never found it and while lambing sheep this year I found their old boxes and dry washer buried in the blow sand which had covered up all trace of work.

The deposit itself was described as a gravel bed three- to twelve-feet thick below an overburden of two feet, starting at river level, extending back about five hundred feet, and which had been thus far proven to a width of ninety feet. Although he indicated that the deposit contained "Good fine gold in tailings," illness had precluded his working the claim, but it was not for sale.

Spencer Canyon

One of the best-known early pioneers in the western Grand Canyon country was jack-of-all-trades Charles Spencer (not to be confused with Charles H. Spencer of Lees Ferry renown). He had the potential to become a prominent miner, but met with an untimely death. Born in England (date unknown), Spencer came to Mohave County, Arizona, in 1863, where he became an army scout, guide, and interpreter. In 1871 he was a guide for the expedition led by George Wheeler, one of the four great surveys of the American West. He was also a mail rider between Hardyville and Prescott. A newspaper account (Arizona Champion, Flagstaff, December 4, 1886) indicates that he was married to a Hualapai woman and had two children, a boy and a girl. Although it appears that he had some troubles with the Hualapai, Spencer was essentially a good friend to them, and they had shown him some samples of silver ore, but were vague as to where they had found it.

In his spare time, Spencer roamed the hills and canyons of the Colorado Plateau and became involved with various mining claims, including a silver mine below the junction of the Little Colorado and Colorado Rivers (Grand Canyon Mining District). He and Dan O'Leary staked a claim near the mouth of the Little Colorado River in December 1877 (Little Colorado River Mining District), but nothing came of it (Weekly Arizona Miner, December 14, 1877). Spencer was also involved with the Ridenour Mine in the Pine Springs Mining District.

Spencer and Charles Cohen were uneasy partners in a ranch near the Hualapai Indian Reservation, and Cohen wanted to sell his interest. Spencer publicly threatened Cohen's life, boasting that he would kill Cohen and throw his body into a canyon so deep that the crows could not fly to it. Some time after the incident, the men got into a fight about ten miles from Hackberry, where the larger Spencer was about to kill Cohen. In self defense, Cohen drew his pocket knife and fatally wounded Spencer. The only witness was Spencer's five-year-old son. Spencer died on November 23, 1886, and on December 7 Cohen was exonerated. During the previous twenty years, Spencer's reputation as an aggressive and reckless desperado was well known, as was his skill with deadly weapons (Arizona Journal-Miner, 1886). The Hualapai threatened to harm Cohen for the killing, but instead named Spencer Canyon, a large tributary to the Colorado River northwest of Peach Springs, Arizona.

In 1968, while in Spencer Canyon, near granite rocks on the west bank of the creek Billingsley found old machinery which he assumed to be left from mining activity. No mine tunnel has ever been found, nor is there any record of who worked this area.

In 1988, Billingsley revisited the site with Jack Antweiler and Karen Wenrich during a U.S. Geological Survey investigation of gold deposits on the Hualapai Indian Reservation. In the nearby granite, they identified hydraulic pumps for washing rock, suggesting a gold placer operation. Several rotting four- and six-inch fire hoses, about fifty feet long, were rolled up nearby. There were scattered remnants of cook stoves and other camp equipment on both sides of the creek. The metal tag on one of the pumps reads "Byron Jackson Iron Wks Inc.," dated August 9, 1921. No claims were found, and the Hualapai do not remember who may have been working there, but they had heard that there once had been some kind of mining venture in Spencer Canyon. Based on the tag date, the work was done after 1921. How the heavy equipment had been carried into Spencer Canyon also is a mystery.

Bat Cave

Not all mining activities in Grand Canyon concerned themselves with gold, copper, and uranium—or even rocks for that matter. Nitrogen-rich bat guano traditionally has been used as a fertilizer, and Grand Canyon has provided this unlikely commodity.

One of the most expensive mining operations ever undertaken in Grand Canyon was at the Bat Cave, deep in the main gorge of the Colorado River at Mile 266, eight hundred feet above the shoreline of Lake Mead. It is located in the north wall, in the Muav Limestone. A young man who had rented a boat for a river trip in the 1930s discovered the cave and passed along the information to Merle Emery. Later, Emery and Beal Masterson of Kingman claimed the cave and tried to mine the guano, but their barges sank in the river.

Emery and Masterson sold the Bat Cave property to the King-Finn Fertilizer Company who tried to remove the

Figure 10. Lower Bat Tower and hopper, just below the Bat Cave. Here guano was loaded into the cable car for transport nearly 10,000 feet across the Canyon to the South Rim (see Figure 11). Photo by George Billingsley, 1976.

Figure 11. Upper Bat Tower and cable headhouse at "Guano Point" on the Hualapai Indian Reservation. Photo by Richard Quartaroli, 1995.

guano by barge and by aircraft. George Steinke of Kingman flew a small piece of earth-moving equipment into the gorge, piece by piece. King-Finn used it to build a small airstrip along the Lake Mead shoreline on a large sandbar that had been created when silt was deposited in the quiet lake waters during a test maximum impoundment behind Hoover Dam. Soon after the airstrip was built, spring flooding of the river washed away the silt.

In 1958, King-Finn sold the property to the U.S. Guano Corporation, a wholly owned subsidiary of the New Pacific Coal and Oil Company of Calgary, Alberta. Frank Ruben of U.S. Guano retained mining engineer Charles Parker to look over the guano deposit. Parker and a nationally recognized firm of engineers estimated that the cave contained 100,000 tons of guano. He also told U.S. Guano officials that the only way to get the guano out of the cave was by an extensive cable system (New York Times, March 27, 1957).

An article in Mohave Magazine (1973) has provided statistics and historical information about the efforts to commercially mine the Bat Cave. U.S. Guano contracted to pay Consolidated Western Steel $439,000 to erect a cable tram from the southern rim of the Canyon just west of Quartermaster Canyon to the mine in the wall on the other side of the Colorado River (figures 10, 11).

During construction, supplies were trucked to Pearce Ferry where Buzzy Wescott of Kingman picked them up with a 1928 Travelair 500 airplane (figure 12). He landed at the reformed sandbar strip at Mile 265. He flew in all the

sand, steel, concrete, food, tools, and machinery (Leonard Pemberton, personal communication, 1984; figure 14).

Steel support towers were built on a point of rock below the South Rim and part way out in the canyon, and on the north side of the river about halfway between the river and the cave. During the construction of the cableway no blasting was allowed at the cave for fear that its roof would collapse. The twenty-seven-foot-deep tower footings for the north side tower were drilled only with jackhammers. The main horizontal span of the cable across the Canyon was 7,500 feet, with a vertical descent of more than 2,500 feet). The cable lines included 9,820 feet of continuous one-and-one-half-inch cable, 20,200 feet of one-and-one-eighth-inch regular plow steel cable for the pulling cable, and one bucket large enough to transport six men and 2,500 pounds of guano (figure 13). For quite unforeseen reasons, this was to become the most expensive aspect of the ill-fated Bat Cave mine.

The first of several accidents and maintenance problems happened when construction of the cableway was nearly completed. The clutch lever broke on the final tension pull on the main cable, dropping the whole cable into the Canyon. It could not be retrieved or reused, so another 9,820-foot cable had to be fabricated in a mill back East.

Figure 12. Buzz Wescott and his 1928 Travelair 500 airplane at the second bat strip (Colorado River Mile 265) in 1958. Photo courtesy of Leonard and Ruth Pemberton, Prescott, Arizona.

Cable construction took fourteen months. Bill Freiday, manager of the guano operation, had a crew of eighteen men working ten hours a day, seven days a week. The guano was sucked from the cave with ten-inch vacuum hoses, stored in a holding bin, transferred to the cable car, and hauled up to the South Rim. There it was dumped into trucks and taken to Kingman where the "high grade" guano, meeting or exceeding the advertised 6 percent nitrogen content, was packaged for sale in one-, five-, ten-, and twenty-five-pound bags, and sent to markets on the west coast.

The Bat Cave Mine provided a brief sensation in the press, probably not so much for its unusual product as for its spectacular means of transportation. Life magazine included some spooky pictures of the operation in its business section (Anonymous, 1957), and the trade journal Good Packaging celebrated both the cable and the Cleveland Container Company cans in which the finished product was sold in stores (Anonymous, 1958).

After four or five months of operation, Freiday, who rode atop the wheel housing every day to inspect the cable, discovered a deteriorating splice in the pull cable. A new sixty-foot section was spliced into the cable, but nevertheless it too began pulling apart, so the whole 20,200-foot pull cable had to be replaced. Consolidated Western Steel fabricated another cable in their eastern mills, and the expensive task of reinstalling a cable was carried out yet again.

About this time in 1958, the cave caught the attention of Paul S. Martin of the University of Arizona, who has studied scientifically valuable plant fragments in cave deposits in the Southwest, principally the dung of large mammals and the urine cemented middens of packrats. These fossils are important in understanding changing climate patterns in the Southwest. Martin made a profile of the Bat Cave with B.C. Arms, principally to plot the positions of three stratigraphic sections of the cave-floor deposits. The profile (figure 15) shows a typically random solution pattern in the Muav Limestone. The bat colony is positioned in the highest point of the cave.

Once the new cable was in place, to help recoup financial losses, U.S. Guano crews moved deeper into the cave and a more concerted effort to mine out the guano was begun. Then a bitter truth was revealed. The cave contained only about 1,000 tons of guano, not the 100,000 tons originally estimated—the remainder of the hoped-for deposit was just decomposed limestone. The little remaining guano was sucked out of the cave in a couple of weeks, and the mine was shut down.

Figure 13. The bat cable car at the Upper Bat Tower on the South Rim. Photo by George Billingsley, 1976.

Several months after the Bat Cave venture was terminated in 1960, a U.S. Air Force jet from Nellis Air Force Base, Nevada, "hot-dogged" down the Canyon (an action then and now prohibited by flight regulations). The pilot was not aware of the cable descending across his path, caught his wing on it, and severed the pull cable. The cable and six to eight inches of the wing tip crashed into the canyon to join the other cables, but the jet returned safely to its base.

After expending three and one-half million dollars, the doomed Bat Cave venture came to a certain and unceremonious end. However, the jet accident precipitated a law suit against the federal government and resulted finally in the guano company recouping the losses incurred in the destruction of the cableway (Huntoon, 1989b).

In February 1961, a Los Angeles public relations agency contacted an agent of the Fred Harvey Company suggesting the possibility for sales of cans of guano in the Fred Harvey restaurants as souvenirs of Grand Canyon: "Since Bat Guano is part of the publicized heritage of Grand Canyon and since it does tend to excite the imagination, I

Figure 14. Aerial view of the bat strip at Mile 265, as it appeared in 1958. View is looking upstream. Photo courtesy of Leonard and Ruth Pemberton, Prescott, Arizona.

There was no evidence of water. The cave was seen to average about twenty feet in width and forty feet in height. It extends an undetermined distance—the examination was abruptly cut short by a massive exodus of disturbed bats.

The Bat Cave was visited by Kenton and Diane Grua in March 1995, and they provided a report on the current condition of the cave and the mining operation (personal communication to Spamer). The one-and-one-half-inch main cable could be seen from the talus chute below the cave. The sheave support and counterweight towers seemed to be sound, although wood planking was rotting. Atop the sheave support tower there were a few rocks, presumably fallen from the cliff. The ladders were in place, and hinges on porthole doors were frozen. A canvas vacuum tube on the terminal tower was shredded and could come loose. The one-and-one-half-inch track cable hung from the tower, attached at its anchor. No remnants of the twenty-thousand-foot continuous pull cable were seen there. On the opposite side of the river, however, the Gruas saw remnants of two cables, presumably the original and replacement track cables.

believe that it would be a popular souvenir item in your Grand Canyon outlets." The guano could be had "at less than half the normal wholesale" price of five dollars per case of twelve one-pound cans, or seven dollars per case of six three-pound cans. More significant, however, was the recommendation that Fred Harvey purchase the tramway, "an extremely attractive tax loss carry-over." The object was to encourage the development of a "resort or tourist attraction," a "Harveyland." A "wonder ride on the tramway would, for sheer dramatic impact, out-rival Disneyland."

In 1973, George and Susan Billingsley examined the Bat Cave. The new population was producing a fresh layer of guano, encrusting the cave floor and the boards and other remaining mining materials. Bats by the thousands hung from the ceiling in chandelier-like bunches.

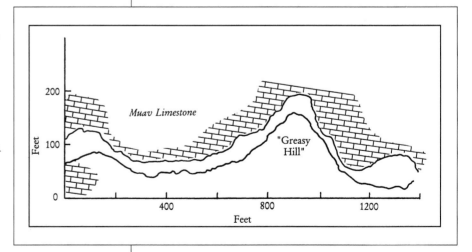

Figure 15. Profile of the Bat Cave, surveyed by Paul S. Martin and B. C. Arms, June 10-11, 1958. The vertical section is exaggerated to show the dissolution pattern in the Muav Limestone. Redrawn with permission from a copy received from Paul S. Martin.

(Richard Quartaroli confirmed in April 1995 that the track cables also remained on the slope between the cave and the river.) Outside of the cave there were "several lengths of 10 inch diameter pipe, several sections of steel ladder, an old winch, and a few other bits and pieces of metal." The cave itself was reported as follows (slightly edited from later correspondence), which agrees well with the conditions observed by the Billingsleys in 1974:

Access to the cave itself is difficult now. The NPS has removed much of the steel ladder going up the 70 foot cliff behind the tower; there is still a section at the base of the cliff (against which leans something that looks like a part of an antenna), as well as a 20 foot section at the top, which is suspended rather dubiously by an old piece of 1/4 inch cable …

Beyond the cliff, there is a talus slope leading to the cave. Midway up it lies another section of steel ladder, which folds back upon itself. A 2 inch pipe and a 5/8 inch cable extend from inside the cave down the talus slope and cliff and past the towers. [The cable passes through an A-frame built of two 3 inch metal pipes beneath the terminal tower, then continues down toward the river.] They may be vestiges of the pre-aerial tram mining efforts. We put on respirators mid-way up the slope; the odor is quite powerful.

At the entrance there is a wooden platform about 20 feet wide by 40 feet long. It is built up about 20 feet off the floor of the cave at the mouth and provide[s] a level working surface. The front is faced with siding and has two 2x4 ladders, still intact, which access the platform. There is a large sheet metal sign on the front which warns "Positively No Smoking." In the center of the platform is a 4x4 foot hole with a 10 inch pipe centered in the opening. Running back into the cave from the platform are a 5/8 inch cable, the 2 inch pipe, and a 2 foot planked walkway about 2 feet wide. The walkway extends about 600-700 feet into the cave, with the cable and pipe located about 6 feet above it. The cable is part of a bucket tram; the tram bucket is back in the big room of the cave at the cable terminus. Also in the big room were a 5 gallon bucket of grease and a 2 foot wide by 5 foot long by 3 foot high machine—we couldn't figure out what it was for. The tram and bucket were part of a pre-US Guano effort. Along the way into the cave and past the tramway on the far end of the big room running up a steep guano-covered slope were several lengths of 10 inch steel pipe, probably part of the vacuum system used by US Guano. Scattered throughout the cave were various odd ends of pipe and other garbage.

When we first entered the cave, we heard the noises of bats—soft, but distinct. The plankway has a thin layer of guano on it. As we walked back into the cave, we passed through portions devoid of bat noise and guano on the planking. Around the end of the planking there were more bats. Continuing past the planking, the guano grew deeper.

At the base of "Greasy Hill," there is a minor passage angling off to the right. Numerous bats were visible. We initially took the left fork, ascending the hill while stepping over the 10 inch pipe. Beyond the pipe we descended the hill, ending in a terminal room in which there were a few bats.

After we returned to the junction, we started up the right fork. Again, there was a hill, but this one was not adorned with mining paraphernalia. The guano was quite deep, and the bats very numerous. Looking down, we could see many bat skeletons on top of the guano, eerie in our headlamps. To avoid disturbing the bats further, we returned.

We later learned from a former boatman who had visited the cave about 15 years ago that there is a small passage beyond where we stopped in the right fork. When he went through it, there was a rope which accessed the next room, the floor of which was about five feet below the passageway. When he and a few friends went down the rope, it broke. Apparently this was the end room, and it did not contain any mining artifacts. In order to return, they had to belly up the five-foot cliff covered with fresh guano. They later disposed of their clothing.

In early 1995, the National Park Service issued a plan for public comment on its intention to have U.S. Navy demolition experts remove the steel towers on the north side of the river, to eradicate trailing on the slopes, and to post the area off-limits to visitation. The park service's object was to better protect the habitat of the Mexican free-tailed bats (*Tadarida brasiliensis*) that live in the cave (Crumbo, 1995). The flurry of new interest in the Bat Cave produced articles in regional newspapers. Mohave County administrators expressed concerns to Fife Symington, governor of Arizona, who in turn discussed the matter with Robert Arnberger, superintendent of Grand Canyon National Park. As of October 1995 the plan to raze the towers on national park land had been set aside.

Today, the Hualapai Tribe is encouraging tourism in this dry area called Grand Canyon West. A few tourists arrive by automobile, but most of them come on air tours that land single-engine airplanes on a dirt strip near

Quartermaster viewpoint. As a focus of the tours, the Hualapai use the cable house on the south side at "Guano Point" (figure 11). Inside they prepare beef with barbecue sauce, vegetables, and cornbread for the tourists, who dine inside the empty remnants of the shed or outside near the rim, but as yet there are no interpretive signs informing visitors about the cave or the guano operation. The Bat Cave can be seen in the Canyon depths, but the cable towers below the mine entrance are more difficult to spot without binoculars. Aside from the cable house, adjacent towers, and the access road, there is nothing left of the Bat Cave operation at Guano Point.

Nelson Mine

Some prospectors in the Grand Canyon region found that some of the very common, high-quality limestone of the region is a valuable commodity. The Redwall Limestone was, and still is, quarried on the Coconino Plateau just off the southeastern corner of the Hualapai Indian Reservation near Nelson, Arizona, a Santa Fe Railroad siding. This operation began in the mid-1890s as the Grand Canyon Lime and Cement Company and continued to operate as the Flintkote Limestone Company. Its proximity to the Atchison, Topeka, and Santa Fe Railroad (today the Santa Fe Southern Pacific Railroad) was certainly a determining factor in the profitability of the venture.

The Nelson quarry, in the Mooney Falls Member of the Redwall Limestone, is in business today as the Gemstar Lime Company, making it the longest-operating mine in the Grand Canyon region. As of 1988, the mine was a large open-pit quarry (figure 16) that utilized both truck and rail to haul the processed cement and road materials to market.

Gold on the Hualapai Indian Reservation

In the spring of 1988, the U.S. Geological Survey carried out a reconnaissance survey of the Hualapai Indian Reservation at the request of the Hualapai Tribe which was determining the resources available to develop their land and economy. The project, headed by Karen Wenrich with several student assistants, set up a base camp at the mouth of Diamond Creek. A helicopter was used to carry personnel and equipment to sampling sites in all of the reservation's canyons south of the Colorado River where Early Proterozoic rocks are exposed, from Granite Park on the east to Quartermaster Canyon on the west. Sampling also included a few localities on Hualapai land along the Grand Wash Cliffs, near Clay Springs and the Music Mountain Mine. A total of 234 samples were collected and panned at the base camp. Just four samples showed any visible gold in the pan, all of them from the Music Mountain Mine area. There was black sand from Tertiary gravel deposits in Milkweed, Hindu, and Peach Springs Canyons, but no visible gold. After laboratory analysis, one sample from a tiny tributary to Diamond Creek was found to contain an anomalous invisible gold content of 558 parts per million. Otherwise, no sample yielded more than 2 parts per million (Wenrich et al., 1989b).

Centennial Mining District

Continuing upstream along the Colorado River from the Grand Wash Cliffs to Diamond Creek, very few mining

Figure 16. Aerial view of the Nelson limestone quarry and Santa Fe Southern Pacific Railroad yard. View is to the west. This is the oldest continuously operating mine in the Grand Canyon region. Photo by George Billingsley, 1985.

ventures are known on the north side of the river; however, north of Diamond Creek there was a discovery of copper in Andrus and Trail Canyons, two areas remote even by today's standards. The mining claims in Andrus Canyon were the original center of the Centennial Mining District around the 1880s, but because the Mount Trumbull Mining District was centered around the Copper Mountain Mine in Parashant Canyon just two miles to the east, we here treat the Andrus and Parashant areas as a part of the Mount Trumbull district. (The latest mining claims at the Snyder Mine and the Old Bonnie Tunnel, on the other hand, have been indicated to be in the Bentley Mining District. This past confusion of districts only emphasizes the uncertainties of the precise locations of these mines.)

Both the Snyder Mine and the Old Bonnie Tunnel are in the Centennial Mining District, but in 1940 they were recorded as being in the Bentley Mining District of Mohave County, Arizona.

Snyder Mine

Many early prospectors did little more than roam around the labyrinth of side canyons that comprise Grand Canyon. Some were quiet about their ventures, and scarcely little was ever known about them. One such character was "Old Man" Snyder, who wandered freely in the remote tributaries of western Grand Canyon.

In 1891, the Snyders, by accounts father and son (the Kingman newspaper said they were Jud Snyder and his brother), wrecked their boat in upper Lodore Canyon, Utah, while on a prospecting and trapping expedition down the Colorado River. Their adventures took them to the western tributaries of Grand Canyon. The Snyders worked a mine, assumed to be in the Trail Canyon area, on the north side of the Colorado River north of Diamond Creek. Old Man Snyder had been mining there for quite some time, digging out copper ore, packing it to the Colorado on his mule, and floating ore and mule across to the south side. There he transferred the ore to a wheeled cart, hitched up the mule, and drove forty-five miles to Hackberry, probably through Peach Springs Canyon. Jud, on the other hand, had come from Utah to hide after killing his sister's violently abusive husband.

In January 1912 Grand Canyon photographic pioneers Emery and Ellsworth Kolb, on their historic filming expedition down the Colorado, came across Old Man Snyder at an unspecified rapid upstream from Diamond Creek (possibly Granite Spring Rapid, Mile 220.5) (Kolb, 1914: 256-257). Clad in tattered garments, he was sitting in a dugout that was about five feet deep and roofed with mesquite wood and an old blanket. Snyder told the brothers how a month before he had come by trail with a burro and a pack mule from Diamond Creek over a one-thousand-foot-high ridge (the east side of Diamond Peak). He had just been out near the top of the Canyon on the opposite (north) side of the river, where he was doing some assessment work on copper claims. He was certain that he would very soon be wealthy. To get to his claims, Snyder had made a raft to cross the river. He said that on one occasion he had been drawn over the rapids, but he managed to get out without injury.

Jimmy Guerrerro, a cowboy for Preston Nutter at Parashant, a small lumber town near Mount Dellenbaugh, gave this account of a person who may have been one of the Snyders (Cox and Russell, 1973:213):

I was coming along a narrow, twisty trail that led under a ledge by a waterhole, and when I stopped to let my horse drink, I heard a kind of noise overhead and looked up. At first, I didn't know who or what it was looking down at me. It didn't have any hat on and hadn't shaved since it was born. I didn't know there was anyone within a thousand miles of me, and then look up to see a mug like that! I tried to pull my six-shooter out; I jerked three times I was so excited; and when I couldn't pull it loose, I looked up again and recognized this Old Snyder. "You old so-and-so," I told him, "don't ever pull a trick like that on me again, or I'll shoot you whether I know who you are or not!" I loaned him a horse and took him over to Parashant and fed him. There used to be a lot of old guys like that prowling about.

The Paiutes knew Old Man Snyder, a Mormon, who would stay with them when he walked to St. George, Utah, to do Temple work. On one occasion, when he did not arrive at the expected time, the Paiutes went to investigate. They found him alone and dead at his mine (Cox and Russell, 1973:213). The mine, known to be in the western Grand Canyon, was for some time afterward the subject of many lost mine stories.

According to the Mohave County Miner (November 14, 1914) and Malach (1975), for a month in the 1920s two men, H.W. Roup and W.H. Davies, searched for the lost Snyder Mine, ostensibly by following the "vein" that was worked at the Grand Gulch Mine. Roup and Davies started their search from Diamond Creek, crossed the river, and found the old Snyder Trail. They were certain they had found the lost mine because they came upon an old mining claim, posted and still legible, with the Snyders' names and a description of the discovery point. So it seemed that the Snyder Mine is very near to the location where Old Man Snyder was prospecting when met by the Kolb brothers in 1912. Roup and Davies found the Snyders' camp and itemized "four cans of condensed milk, about one pound of dried apples, a little flour, a little baking powder, a little salt and a few beans" (Miner), actually an incredible inventory considering the voracious appetites of small animals like packrats. The newspaper reported that the men had found the mine and recovered copper ore assaying at an average 65 percent and up to 85 percent copper.

The last claims on the Snyder Mine property were dated October 15, 1958, by Robert Williforb of St. George and Glen Summerlin of San Fernando, California. They named the mine Futile.

While doing reconnaissance mapping in western Grand Canyon in November 1975, George and Susan Billingsley found a well-constructed old trail which extended from near the Colorado River to the rim of the Redwall Limestone; it led straight to the Snyder Mine. The mine is located in a breccia pipe in the Watahomigi Formation just above the Redwall Limestone. When found, it was in good condition and contained a rich deposit of copper minerals such as chalcocite, azurite, and malachite. Various iron minerals were abundant, too. Chemical analyses of the ore show anomalously high concentrations of arsenic, cobalt, nickel, molybdenum, and zinc (Karen Wenrich, personal communication, 1989). Several old mining tools and other assorted equipment, such as a wheelbarrow, were found there.

Veteran Grand Canyon traveler Harvey Butchart found another trail that leads from the Colorado River to the Snyder Mine. The foundation for a miners cabin or tent can be found near the Colorado River. Several old tin cans, some purple glass, and a rusted pocket watch with Roman numerals on the face were found at the site by Billingsley in 1976.

Old Bonnie Tunnel

Virtually nothing is known about another prospect called the Old Bonnie Tunnel on the USGS 7.5 minute Granite Park quadrangle. Its proximity to the Snyder Mine suggests that they were discovered at the same time. However, in 1975 George and Susan Billingsley inspected the tunnel and found a claim note dated June 5, 1940, by H. Loraine Cox, Lee Cox, Roy Wood, and Perry Wood, which may be the first recorded claim. The name Copper Flat was on the

claim, with no mention of the Old Bonnie Tunnel. In 1941 correspondence with Charles F. Willis, of the Arizona Department of Mineral Resources, there is an undated handwritten response from Louis Arnold of St. George regarding the ownership of mines in northwestern Arizona. The "Snyder" mine and the "Copper Flats" mine were indicated to be owned by the Coxes and Woods; the Coxes in the care of the Arrowhead Hotel in St. George, and the Woods at Post Office Box 411.

The mineral assemblage in the Old Bonnie Tunnel is similar to that of the Snyder Mine. It is located in a small breccia pipe in the Redwall Limestone, nearly overlooking the Colorado River. Samples collected from the tunnel by Karen Wenrich in 1983 showed an even larger suite of anomalous metals than those in the Snyder Mine—arsenic, silver, cobalt, mercury, nickel, molybdenum, lead, and zinc (Wenrich, personal communication, 1989).

Mount Trumbull Mining District

While gold-seeking prospectors and cattlemen ventured into canyons from Diamond Creek to the Grand Wash Cliffs, others came from the north looking for silver in the red sandstones of the Pennsylvanian and Permian strata in Grand Canyon, which were natural attractions with great exposures. Only copper was found, although the Copper Mountain Mine in Parashant Canyon and the Little Chicken Mine near Cove Canyon are the only two breccia pipe structures of hundreds on the Colorado Plateau where gold is reported from chemical analyses (Karen Wenrich, written communication, 1989).

In the eastern part of the Mount Trumbull Mining District, there are several interesting mining claims in and around the Toroweap Valley area (Tuweep on older maps). Although this area was never in any proclaimed mining district, its proximity to Mount Trumbull allows us to declare here that the claims in that area are part of the Mount Trumbull Mining District.

North of the boundary of the Mount Trumbull Mining District as depicted on Map 2, but which probably should be included herein, Billingsley discovered in 1994 what appears to be a copper prospect in a stratiform breccia deposit (not a breccia pipe). The mine is about twenty miles north of Whitmore Canyon, in the Dutchman Draw 7.5 minute quadrangle. It is not much more than a prospect pit, and there is little evidence of any copper production from the small shaft about six feet deep. Copper minerals (malachite and azurite) are present in a sandy limestone bed of the middle Harrisburg Member of the Kaibab Formation.

Nothing remains of the miners' operation except some broken purple glass and rusty tin cans. It has the general appearance of having been worked around the 1940s. No subsequent prospecting work seems to have been done at this mine, which was discovered when Billingsley was looking for a campsite out of the wind.

Copper Mountain Mine

By 1890, prospectors had explored virtually every side canyon in Grand Canyon. One interesting prospect that seemed to pay off for a while was the Copper Mountain Mine. The prospect was located April 10, 1875, on a breccia pipe in the Esplanade Sandstone southwest of Mount Trumbull. This mine is extremely isolated, even today. A one-hundred-mile trip along very poor dirt roads is necessary to reach the mine from the nearest settlement, St. George. The last dozen miles, down Mule Canyon, can take three hours on a good day.

The Copper Mountain Mining Company was composed of four unpatented claims over about twenty acres. It was incorporated on May 19, 1875, in St. George. By February 21, 1881, the mine was reassigned to Mohave County, Arizona, from Washington County, Utah, by A.P. Hardy, E.G. Wooley, Robert C. Lund, James Andrus, E.D. Wooley Jr., and E.W. Snow. It was recorded at the request of James Andrus, president of the Copper Mountain Mining Company, and Edwin G. Wooley, secretary. Copper was the commodity taken from the mine.

John A. Swapp of Overton, Nevada, owned the mine sometime between 1890 and 1914. The mine operator then was Bishop Whitehead, also of Overton (Hill, 1913).

According to reports from Pat Bundy of Bundyville (Mount Trumbull), Arizona, Reed Mathis of St. George, and Laura Gentry of St. Thomas and Overton, Nevada, around 1915 the mine owner was an elderly woman remembered only as McMasters. Belshaw's (1983) report cites that every two or three months from about 1915 to 1917 ore, containing 23 percent to 26 percent copper and yielding three to four dollars of silver per ton, was sent from the mine to smelters. According to Belshaw (1983:9):

The road going to Copper Mountain went by way of Pakoon and passed another cow camp called Hidden Father along the trail. The road came up on top of the mountain a good deal farther north than the Grand Gulch Trail. The Copper Mountain mine was down in a deep rugged canyon and the ore was packed out on burros for seven miles to where it was loaded on the freight wagons.

At times there were two or three hundred burros in the pack train. It was quite a sight to see them strung out on the trail. The ore from this property was very high grade, as nothing but the best of ore could be shipped hauling it from that distance.

According to Cox (1982:313), Ivy Stratton, who worked at the sawmill at Parashant on the Shivwits Plateau, and his brother hauled lumber to the Grand Gulch Mine in 1909. In 1914, Ivy went to work for a few months at the Copper Mountain Mine. He recalled, "It had good ore, but the only way to get it out where teams from St. Thomas could pick it up was by using a train of about 25 burros. The pickup point was near Andrus Spring at the turn of Parashant Wash, and they would have to go down through Pigeon [Canyon] and out by way of the Grand Gulch." Some time later, a wagon road was built up through Andrus Canyon to the Shivwits Plateau. According to LaDell Jessup, who also worked from time to time at the Copper Mountain, the road down Andrus Canyon washed out in a storm (probably during the late 1950s) (Cox, 1982:314).

Having walked down this road to the mine, Billingsley is certain much of the miners' time must have been occupied just maintaining the road to be passable by horse, even more for wagon or truck. Today, the road is nearly erased from the canyon narrows, although green copper ore scattered along the drainage marks the old trail.

In 1941, correspondence from Louis Arnold of St. George, to Charles F. Willis, of the Arizona Department of Mineral Resources, indicated that the Copper Mountain Mine was owned by H. Loraine Cox, Lee Cox, and Etta Cox, with addresses in care of the Arrowhead Hotel in St. George. The Copper Mountain Mine was intermittently active in the late 1940s and early 1950s, when a new road was built down Parashant Canyon and an airstrip was laid out. LaDell Jessup tells of the time his brother, Fenn, and Merlin Peterson built an airstrip a mile west of the mine on the Esplanade Sandstone by dragging weighted bedsprings around behind a pickup truck. "Then they flew three mine rails at a time underneath a Super Cub from Grand Gulch down to the Copper Mountain. They would make three or four trips a day" (Cox, 1982:315).

After World War II, Western Gold, owned by Leland Whitmore Sr., attempted to work the ore body for gold and uranium. Western Gold and Uranium, Inc., also examined the mine but never drilled or did any work (William Chenoweth, personal communication, 1987). The Nuclear Exchange Corporation (predecessor of the Atomic Energy

Figure 17. Hoist at the Copper Mountain Mine, as it appeared in 1984. Photo courtesy of Robert C. Euler.

pipes in the Parashant Canyon area, including the Copper Mountain, Copper House, and Chapel Mines, but the company never mined any of those properties. Much of the mining equipment of previous years still remained when Billingsley visited the mine in 1968. Robert Euler (personal communication) also reported that equipment remained when he visited there in 1984 (figure 17). Foord et al. (1978) reported that the Copper Mountain Mine produced copper, silver, gold, lead, and zinc (see Table 1).

The Copper Mountain Mine derives what interest it might have not from its production, but from creating problems years later for the National Park Service. Some of the hundreds of burros used to haul ore from the mine escaped and found their way into the remote reaches of Grand Canyon, where they lived in isolation and maintained a lively population that spread along the banks of the Colorado River. Together with burros on the south side of the river—and the burros presumably swam across the river during lower flows—they upset the natural ecological habitat in the river corridor, competing with bighorn sheep and other animals that depended on the same sparsely available foods. The burros were eventually rounded up and removed by the park service and animal interest groups in 1977 and 1978.

Chapel House Mine

On April 10, 1875, Daniel Seegmiller and E.W. McIntyre located a mine they called the Copper Head; they refiled the claim on February 22, 1881. This mine may be the same prospect now called the Chapel House Mine (also known as the Bundy Prospect), in Mule Canyon. The area around the adit was claimed in 1952 by C.M. Bundy of Mount Trumbull, and Omer Bundy of St. George. On February 24, 1954, C.M. Bundy shipped 1.08 tons of ore averaging .23 percent uranium oxide, 4.02 percent copper, and 1.10 percent calcium carbonate to the Atomic Energy Commission ore buying station at Marysvale, Utah (Chenoweth, 1988). The mine area was held by Mrs. Audrine C. Knight under a bonded lease with the Bureau of Land Management and the National Park Service, approved on April 1, 1966. Foord

Commission) (1978) reported that the Copper Mountain Mine produced seventy-five pounds of uranium oxide ore in the 1950s, but could locate no record of this production in its files (Chenoweth, 1988).

The mine dumps were reworked intermittently from 1955 to 1961 by the Signal Oil and Gas Company (Cox, 1982:314). In 1964, claims were included in Lake Mead National Recreation Area and are unpatented. The property was last owned as a uranium prospect by Exxon Minerals during the mid-1970s. Exxon drilled many of the breccia et al. (1978) reported that the "Chapel prospect" had shipped only "a few tons" of copper.

The Chapel House Mine is in a poorly exposed breccia pipe that crops out in the Hermit Shale. It is partly hidden by extensive talus debris on the west side of Parashant Wash, about three hundred feet below the Coconino Sandstone. The breccia pipe is estimated to be at least one hundred feet in diameter. A fifty-five-foot-long adit was driven southward along a fracture zone in a bleached, fine-grained sandstone. Nelson and Rambosek (1970) noted uranophane, autunite, and copper oxides at the mine. Minerals reported by Denys K. Poyner in 1972 (Atomic Energy Commission files) are copper carbonates, malachite, chalcocite, azurite, and unidentified silver and secondary uranium oxide minerals. From a sample of the copper ore, Fukui (1982) identified azurite, chalcocite, covellite, and brochantite.

Five drill holes were cut into the breccia pipe by the Cotter Corporation in 1968. Cotter has subleased the pipe from Mrs. Knight. Although anomalous radioactivity is present, and some copper mineralization occurs both at the surface and at depth, no economic uranium has been indicated (National Park Service, 1977).

The lease was transferred to Exxon Minerals, USA, on November 1, 1974. Exxon had planned to drill into the pipe, but permission was not granted by the Bureau of Land Management and the National Park Service. A renewal of the lease was denied on December 9, 1978 (Chenoweth, 1988). Exxon discovered uranium minerals such as autunite, torbernite, uranophane, and pitchblende, disseminated through the rock and along fractures and in veins. Their chemical analyses indicated an average of 0.20 percent uranium oxide, and about 9 percent silver and copper (National Park Service, 1977).

Other Claims in the Parashant Canyon Area

Another placer, known as the Johnson placer claim, was located on June 4, 1893, at the mouth of Parashant Canyon in western Grand Canyon, by J.S. Johnson, W.M. Johnson, H.E. Judd, S.N. Johnson, J.B. Francis, Joe Meeks, C.A. Huntington, and Jeremiah Johnson. Again, dreams of riches ended when little gold was produced.

Several claims for copper were located during November 1880 in the area now known as Whitmore Canyon, by B.M. Ellenbeck, C.A. Smith, John H. Cassidy, John Quillan, Mat O. Loughlin, and John H. Rice. Trails were built to the Colorado River, mainly to get water for livestock, miners, and ranchers. Little or no mining was ever done on these claims. The trails are still in use today, for

hiking access to the Colorado River and for taking out some of the Colorado River rafting parties.

Claims were also made west and south of Parashant Canyon, where the Centennial Mining District was established. W.G. Chidester and J.D.L. Pearce filed several claims for copper in May 1883, but none of the claims are known to have produced economically viable ore. The Copper House claim in Dansill Canyon, located on a breccia pipe in the Hermit Shale near Mud Springs, showed promise of becoming a good copper producer, but this never came about. A few of the minerals recognized during visits to the mine in 1976 by Billingsley were malachite, azurite, chrysocolla, chalcocite, limonite, and hematite. Wayne McConkie gave this account of the mining equipment found near the mouth of Parashant Canyon (Cox, 1982:152):

I went with [Norman] Nevills on an expedition down [the Colorado River] through the Grand Canyon in 1940. And way up under an overhang, or ledge, . . . we found some mining equipment and other stuff that had probably been there for thirty-one years! It was an extremely well-protected spot. There was a rope, still coiled, a pick and shovel, and a frying pan with a can of Eagle Brand Condensed Milk sitting in it. And the date on that can of milk was the year of my birth, 1909! Everything seemed to be in perfect condition, and I have wished many times since then that I had slipped that can of milk into my duffel bag. Everything has probably been taken long ago by someone else doing a bit of exploring up in those ledges. I have no idea who the prospector was that left the stuff there. He probably came into the country with burros and left them there somewhere while he climbed around prospecting. And we were quite likely the first ones to find his cache. I doubt that he ever returned to it after placing it there.

By the mid-1970s, most of the claims in the Parashant area were acquired by Exxon Minerals, USA as uranium prospects. Exploration holes were drilled, and some uranium oxide was found at the Copper House and Copper Mountain mines, but the claims expired without production. A new boundary for Grand Canyon National Park was established in 1975, but because of the numerous mining claims in this area it was not included in the park. The area is still being considered for inclusion in the national park, but there are no concrete plans at this time. Uranium companies hold some of the old claims for possible use in the future.

Foord et al. (1978) reported that the "Copper House Colition [sic] Nos. 1 and 2," although containing copper and uranium commodities, had produced only copper. A prospect identified in sec. 31, T. 33 N., R. 9 W., containing copper, had no production.

Little Chicken Mine

One area of mining interest in this district is located near Toroweap Valley north of the Colorado River. Local ranchers in the valley relayed to John Riffey, park ranger in the former Grand Canyon National Monument, area mining stories. John wrote down some of the information he had heard and believed to be true, including some details of mining activity.

Brady Inglestead, "Windy Jim," located a copper and silver mine below the rim at the top of the Redwall Limestone (figure 18). The exact date of location is not known, but on August 16, 1921, the mine was called Little Chicken (Mohave County Courthouse, book of mines 3-B, page 452). Wilfred Brooksby, as a boy, worked the mine for Windy Jim. A windlass was built to lower supplies and haul up the ore in a sling over the five-hundred-foot-high cliff of Esplanade Sandstone above the mine. Windy Jim had a partner in the venture, Henry E. Covington, whose name he forgot to file with the claim. Covington made sure his name was on later claims, and on April 25, 1929, Covington was the only one to refile on the claim, changing the name to the Ram, and Shepherds Folly on December 10, 1938. The U.S. Department of Interior declared the claims null and void in the late 1950s, but Covington never accepted the declaration (U.S. District Court of Arizona, 1960).

John Riffey said that Covington's crew hauled out to Pueblo, Colorado, some twenty to forty tons of ore, much of it containing silver and lead sulfides similar to those found in Havasu Canyon (see Havasu Canyon Mining District). In 1961, much to Riffey's displeasure, a road was built on the Esplanade to the rim above the mine. The National Park Service rose to the occasion and fought the development as a negative impact on the wilderness of the Arizona Strip. The improvements were declared null and void by administrative decision, November 19, 1970, and ownership reverted to the national park.

In the summer of 1985, Karen Wenrich of the U.S. Geological Survey visited the Little Chicken Mine to collect ore samples for chemical analysis and determine whether rumors of gold there were true. At the time, Wenrich and Billingsley determined that the mine was located on a breccia pipe in the Mooney Falls Member of the Redwall Limestone. The chemical analysis did indeed show traces of gold, making it only the second breccia pipe on the Colorado Plateau known to contain any. The analysis also showed very high concentrations of zinc and lead, with lesser amounts of copper, arsenic, barium, cadmium, and silver (Wenrich, personal communication, 1988).

Most, if not all, of the mining tools were still in the mine in 1985. A miner's camp is located in the vicinity of the mine. In addition, a kiln (perhaps only a fireplace) can be seen on the Esplanade rim near the windlass, but its relationship to the Little Chicken Mine is unclear and its builders unknown.

Brady Canyon Claim—A Water Mine

John Riffey wrote of another interesting mining adventure about which he heard from Marcell Schmutz, a local rancher, on March 18, 1941. In Brady Canyon, a tributary on the east side of Toroweap Valley a few miles north of the Colorado River, there is a hole that looks like a mine shaft. Schmutz recalled:

During the spring of 1925 I was in Tuweep Valley hunting horses and late in the evening was coming by the mouth of Brady Canyon and saw a campfire up in the canyon. When I came closer I saw three men around the fire, looked like Indians but were white men. They were Brady Inglestead, Frank Heaton, and another fellow. I camped with them that night and they told me the following story. They were digging for water up there and 4 hours out of 24 they could hear water flowing back in the rock. In a hill running east and west they could hear it running in a westerly direction. They thought the Indians had dammed it off, since there were over 100 year old Indian houses in the valley. They figured the Indians had watered the valley with it then hid it when they left. They figured to run a tunnel in and catch it where they could hear the water running. It didn't run at set times but just any time about 4 hours a day. They ran a tunnel in about 254 feet, didn't hit water and finally gave up for lack of money and ambition to go any deeper.

The tunnel is still there, mostly filled with debris.

Tuckup Canyon Mine

Farther east from Toroweap Canyon is a large tributary called Tuckup Canyon (also known as Tucket), included in the former Grand Canyon National Monument and since 1974 in the national park. During a geological

Figure 18. Aerial view of the Little Chicken Mine (arrow) at the top of the Redwall Limestone. The broken slopes below the mine are remnants of an ancient landslide. Photo by George Billingsley, 1985.

reconnaissance by Billingsley in 1968 and 1969, an old abandoned copper mine was found on a small brown knoll near Cottonwood Spring. John Riffey wrote a few notes on this prospect, based on information he obtained from Marcell Schmutz, a local rancher.

The Tuckup Canyon prospect was first worked about 1911 by Joe Price and a man named Hunsucker. It was worked by Marcell Schmutz and his father around 1921. Schmutz said, "Several other fellows have jumped the claim since 1921 but haven't done any work. The ore contains copper, gold, silver, and lead. It was too long a pack out for it to be a money maker." Billingsley found a shaft that was sunk about seventy-five feet, and two short adits dug into a hill nearby. A wheelbarrow and several tools were still in one adit in 1970. The copper minerals azurite and malachite lay around, but there was nothing that looked like it would contain gold.

Two of Schmutz's "claim jumpers" were Walt and Rasco Cumingham, who laid claim to the mine on September 5, 1936, naming it the Pinto Mine. The name may have come from the spots of blue and green copper minerals in the Esplanade Sandstone. The claim changed hands to Wallace Blake on March 15, 1937. In 1938, William Cumingham took over the property and renamed it the Grand Canyon Copper Mine. It was abandoned until May 5, 1955, when Jense McCormick of Fredonia, Arizona, took it over. McCormick made several other claims in the area in February 1959, but little if any work was done.

It is interesting that this mine may never have been much of a producer because it appears to be located in a slump block resting on the Esplanade Sandstone rather than in a breccia pipe, making it a very limited deposit. The rocks of the slump resemble the Toroweap Formation in the upper cliffs of the canyon. The shaft was dug into the Esplanade Sandstone below the ore-bearing slump block. Today, geologists suspect, based on the mineral assemblage and isolated locality of the mine, that a breccia pipe may be there. The only way to verify this is to drill the area, and, as it is in a national park, this is not likely to be done.

The Tuckup Canyon Mine may have changed hands so many times because of the ill effects of the only spring near the mine. The water in Cottonwood Spring is good to drink if one does not mind the flat taste of alkali and magnesium. A few good swallows will quench the thirst, but the high salt and gypsum content make the drinker take more. The ultimate effect is purgation.

Pine Springs Mining District

South and east of the Colorado River, bordering the Centennial and Mount Trumbull Mining Districts, are canyons and high plateaus of the northeastern part of the Hualapai Indian Reservation. These areas were not overlooked by early prospectors or by the interventionist activities of George M. Wheeler in the guise of federal mapping surveys (see Music Mountain and Lost Basin Mining Districts).

Ridenour Mine

In the 1870s and 1880s, the remote section of Grand Canyon south of the Colorado River and west of Havasu Canyon saw a flurry of mining activity on the red benchland of the Esplanade Sandstone. William B. Ridenour and S. Crozier roamed through the canyons south of Whitmore Canyon, on land that today is the northern part of the Hualapai Indian Reservation. They said that they had been driven out by Indians in 1874, but they returned later with partners John Tillman, William H. Hardy, and Charles Spencer. Together they located the "Grand Canyon" copper and silver mine on March 6, 1880, but the name was changed to the Ridenour Mine on April 19, 1882.

Charles Spencer, with William Ridenour, the discoverer of the Hackberry Mine near Hackberry, prospected the Grand Canyon area north of Peach Springs in 1883. A letter signed by "Silex," printed in Flagstaff's Arizona Champion on December 19, 1883, mentioned in part, "Ridenour and Spencer feel rich over their copper discovery, and we hear a Mr. Downer from the Black Hills is soon to purchase an interest." On June 16, 1884, Tucson's Arizona Daily Star printed another item:

A mine known as the Copper King, owned by Messrs. Spencer and Ridenour, located in the Grand Canyon of the Colorado, about 60 miles from Peach Springs, has been examined in the interests of California capitalists, and the report was so favorable, the ore by assay yielding from 40 to 80 percent copper and $15 to $16 silver, that they intend to develop the same. Some of the work has been done by the owners and now men are employed in building a road to the property. It is said the deposit was originally found by John D. Lee some 12 years ago, when he was fleeing from justice on account of the Mountain Meadow massacre. In it, has been found a ledge of copper and silver 650 feet in length and from 20 to 60 feet in width.

Foord et al. (1978) indicated that the Copper King, in Hack Canyon, in sec. 26, T. 37 N., R. 5 W., had shipped ninety-five tons of copper in 1890 and 1891, and "8 cars up to 1947."

Ridenour and J.W. Porter also had located another prospect on October 28, 1892, at end of a long ridge in the Hermit Shale then called Lime Ridge (no name today). It is located on a prominent breccia pipe about a mile east of the Ridenour Mine. Although much exploratory work has been done at this prospect, no mining ever took place. Even uranium explorations in the early 1970s failed to find economically viable deposits.

The Ridenour Mine, also called the Zimmerman Mine, was the most productive of William Ridenour's prospects. Buildings or other structures related to the mine are, however, nearly absent. A stone chimney was built near the mine, but it may have been for a tent since there is no evidence of stone walls.

The Daily Arizona Miner of April 7, 1880, carried an article stating that ore from the Ridenour Mine contained from 60 percent to 90 percent copper, and pieces of nearly pure copper metal weighing at least a ton were not rare. The mine workings closely followed the boundary of a circular breccia pipe in the Esplanade Sandstone. It was worked intermittently for copper prior to and during World War I, with most of the activity taking place between 1916 and 1918 when it was still owned by Ridenour. An estimated one thousand tons of copper ore were carried by burro to the canyon rim, then transported by wagon to a railroad siding called Nelson, a few miles east of Peach Springs. Copper minerals identified by Miller (1954b) are azurite, malachite, and chalcocite, with small amounts of chrysocolla and bornite. Uranium and vanadium oxide ores composed of yellow tyuyamunite and green volborthite form a thin crust or shell around the ore body (Bradley S. Van Gosen and Hoyt B. Sutphin, written communication, 1989).

The property has been abandoned since 1920. Miller (1954a, b) reported that in 1953 he made a reconnaissance for uranium on the Hualapai Indian Reservation and produced a detailed study of the Ridenour Mine for the Atomic Energy Commission. He indicated that the mine was under the control of the Hualapai Tribe and not leased. By August 1955, the Atomic Energy Commission was contacted by Sawyer Exploration Company of Los Angeles, California, who said that they controlled the mine and requested federal government assistance to develop it. The company built a small airstrip southwest of the mine and improved the road to the old workings. The U.S. Bureau of Indian Affairs informed the Atomic Energy Commission that Sawyer Exploration did not have a valid lease and that they were trespassing on the Hualapai land (Chenoweth, 1988).

In 1959, Western Gold and Uranium obtained an exploration permit from the Hualapai Tribe and improved the trail to the mine and drilled two core holes. The results of the exploration were negative for uranium, and copper concentrations were seen to be of low grade.

Clyde Hutcheson of Flagstaff signed a ten-year lease with the Hualapai Tribe on May 21, 1960, for 76.8 acres surrounding the Ridenour Mine. In January 1961, Hutcheson made a 14.14-ton shipment of uranium ore to the mill at Shiprock, New Mexico, yielding an average .15 percent uranium oxide and 2.36 percent vanadium oxide (Chenoweth, 1988). The long haulage distance and the low grade of the uranium ore made the effort hardly worthwhile. Another ten-ton shipment of copper ore was sent to the smelter at Superior, Arizona, which yielded 8 percent copper. The two shipments were the only ones made by Hutcheson, who canceled the lease in 1963 or 1964 (Chenoweth, 1988).

The mine was again leased and drilled in 1976 by Western Nuclear, Inc. to assess the uranium potential deeper in the breccia pipe. Three holes were drilled to depths ranging from 324 to 990 feet, disclosing little mineralized rock. In 1978, the company's lease was dropped, and control of the property reverted to the Hualapai Tribe (Verbeek et al., 1988).

National Canyon and Laguna Lake Claims

National Canyon, a large southern tributary to the Colorado River on the Hualapai Indian Reservation, has been well prospected. A small spring in the Redwall Limestone in the lower reaches of the canyon no doubt was an attraction for early prospectors. Access to the Colorado River in this canyon can be made only on foot, not even with a burro.

In 1891, many prospects were claimed in National Canyon and the next canyon westward, Mohawk Canyon (sometimes called Moho Canyon). Most of the copper

claims were filed by J.A. Healy, S. Gilroy, and J.J. Phillips. The National Canyon claims are in a small western side canyon of National Canyon that the prospectors called Boulder Canyon. The claims are situated fairly close together on two breccia pipes in the Esplanade Sandstone. Very little copper ore is found in these pipes, and apparently little of that was mined because the diggings are very small. J.A. Daggs, A. Humphreys, W. Latchford, and Millard Love also had several claims in National Canyon near the same area, but these were mostly just wishful thinking.

In 1986, during the U.S. Geological Survey's breccia pipe study on the Hualapai Indian Reservation, Billingsley found a small, obscure breccia pipe in an eastern tributary of National Canyon. The copper minerals malachite and azurite were seen, but there was no indication that any prospecting or mining had ever been done nor were any claims found. This is one of the few Grand Canyon localities with visible copper minerals that prospectors may have missed.

Farther south in the National Canyon drainage basin, on the Coconino Plateau and about four miles north of the Havasupai road that crosses the drainage, a copper deposit was located in the Harrisburg Member of the Kaibab Formation. The group of claims, known as National Nos. 1, 2, 3, and 4, were located on November 1, 1919, by A.M. Boss and Mack Tokespeta in what was called the Walapai Indian Reservation Mining District (now included in the Pine Springs Mining District). During a 1986 visit by Billingsley and Karen Wenrich, copper minerals were seen as thin veinlets and stringers in joints and bedding planes in stratiform brecciated limestone beds. It is not known how much more was taken from these claims, but judging by the amount of digging in open pits and trenches, probably only small tonnages were hauled out. These deposits are similar to those at Laguna Lake to the west, the Anita deposits to the east (Francis Mining District), and the Jacob Lake area on the Kaibab Plateau north of Grand Canyon (Warm Springs Mining District).

During the U.S. Geological Survey breccia pipe study by Karen Wenrich and others, between 1983 and 1987 seven small and mostly forgotten copper prospects were found on the Aubrey Cliffs and in Prospect Canyon northwest of Laguna Lake. Each prospect is little more than a bulldozer cut in the limestones of the Harrisburg Member of the Kaibab Formation. Very little information is known about the Laguna Lake prospects. One of them, located by Charles H. Dunning and T.E. Carger in November 1919, was designated as the Lagoon Nos. 1, 2, 3, 4, 5, 6, 7, and 8 lodes. It is doubtful that much, if any, ore was removed from these sparse deposits. Several Atomic Energy Commission Preliminary Uranium Reconnaissance Reports from the 1950s describe some of these occurrences, but the locations and directions in the reports are in error by over a mile.

One report, by Hewett et al. (1936), states that the only recorded production from the disseminated copper ore in Paleozoic sedimentary rocks (Kaibab Formation) in the Walapai Mining District (Pine Springs Mining District) was made in 1929, thirty-seven miles north of the railroad siding at Pica, in the Aubrey Valley. The shipments, valued at more than five thousand dollars, consisted of 183 tons of ore containing copper and small amounts of gold and silver. Probably most of the ore came from the workings near the head of National Canyon on the Coconino Plateau.

Aubrey Cliffs Manganese Prospect

Farther south of National Canyon and north of Seligman, Arizona, and just east of the Hualapai Indian Reservation boundary at the Aubrey Cliffs, a small manganese prospect is located in the Kaibab Formation. Small mine tunnels were driven into a brecciated chert zone, where black manganese fills voids in the chert layers. This has been identified as the Johnson and Hayden deposit, located on land owned by the State of Arizona. Farnham and Stewart (1958:12-13) have provided the following information:

The deposit was discovered and slightly explored in 1940 by Don C. Adams and W.J.E. Woody under a prospecting permit issued by the State of Arizona. So far as known, no ore was marketed while the property was held by Adams and Woody, and their permit eventually was allowed to lapse. Interest in the deposit was revived in 1952 and 1953 after establishment of the Government depots at Deming, N. Mex., and Wenden, Ariz., for the purchase of low-grade manganese ore. In that year E.H. Johnson and R.E. Hayden, of Seligman, obtained a mineral lease on the property and shipped about 37 tons of sorted ore containing 28 percent manganese to the Deming, N. Mex., purchase depot. Later, the property was subleased to the Bosley Mining Co., which during 1953 produced approximately 275 tons of sorted ore averaging about 24 percent manganese. This ore was shipped to the Government purchase depot at Wenden, Ariz. When the area was visited in August 1954, two men employed by the Bosley Mining Co. were producing small quantities of ore from the deposit.

Manganese mineralization on the property occurs within a steeply dipping fracture or brecciated zone cutting the gently dipping beds of the Kaibab limestone. The fracture zone on the surface ranges from ten to thirty feet in width, strikes northeast, and is exposed in places along the strike for over 1,000 feet. Although manganese mineralization was evident in several places along the outcrop of the brecciated zone, the better mineralized portion appeared to be limited approximately to a strike length of 300 feet and an average width of 25 feet. The principal workings were concentrated to this area; they consisted of a centrally located crosscut adit with two short drifts and several pits on each side of the adit. The deepest of these openings was about 40 feet below the outcrop.

Farnham and Stewart (1958) indicated that the minerals were "common oxides," and the mining procedures were simply to "gouge out" the higher-grade masses when they were encountered.

Lost Turquoise Mine

During the U.S. Geological Survey breccia pipe project in 1983, Karen Wenrich and Hoyt Sutphin discovered and named the Mohawk Canyon breccia pipe on the west rim of Mohawk Canyon. Their helicopter pilot, Mike Bertoldi, found an old mine tunnel containing copper minerals on the rim of the canyon, on the edge of the breccia pipe. They found out later from the Hualapai Tribe that this mine must be the Lost Turquoise Mine rumored to be in the Mohawk Canyon area. A rotting windlass is at the front of the mine tunnel. The tunnel follows a ring fracture of the breccia pipe in limestones of the Harrisburg Member of the Kaibab Formation. Malachite and azurite were seen in the ring fractures of the pipe.

The USGS drilled five exploration holes in the Mohawk Canyon pipe. Uranium minerals were detected, but not in the expected quantities (Wenrich et al., 1989a). It has good potential for future development, however, if uranium prices rise. One of the major costs of the drilling project was finding and hauling water to the drill site.

Platinum and Mineral Water

On April 16, 1904, the Mohave County Miner published an account that platinum had been discovered in the western Grand Canyon. While the report probably came from dubious prospects in both the Pine Springs and Music Mountain Mining Districts, the fact that it also pinpoints a singular occurrence of mineralized water permits us to mention it in the Pine Springs district. The newspaper reported:

Col. Wilbur, a mining man of northern California, Jack Cuish and Robert Rosco [and others] returned from a thirty days prospecting trip into the Grand Canyon country, where it is stated they found great deposits of platinum ore. The strata carrying the metal lies in a horizontal strata [sic] in the sandstone cliffs and in places is fully six feet in thickness. A large amount of samples have been taken to San Francisco and should it come up to expectation work on a large scale will be begun on the properties at once. The party found a great spring of mineral water boiling up out of the edge of the river, a few miles east of Diamond Canyon. This spring they believe to have great medicinal properties, and should the mines be found rich in platinum the same company will probably utilize the waters profitably.

The spring is no doubt Pumpkin Spring, at the east side of Colorado River Mile 212.9. In this unique feature, a mineralized spring has deposited an orange-colored half circle of travertine on the Tapeats Sandstone cliff at the edge of the river. Water trickles over the lip into the river, and at low river flows the irregular, stalactite-like lower part of the travertine rind can be seen clearly. The water is warm (86 degrees) and has a sulfurous smell. The bottom of the three-foot-deep spring pool is anoxic and the black mud is easily stirred up. It is good that the spring was never developed. The water is extremely toxic, containing very high concentrations of arsenic and zinc. Contact should be completely avoided (Mazzu and Rihs, n.d.).

Havasu Canyon Mining District

The area around Cataract Canyon and its lower reach, Havasu Canyon, a major southern tributary to the Colorado River on the Coconino Plateau, saw some of the most energetic and diversified early prospecting in the Grand Canyon region. This side canyon plays a major geographic role as a convenient access to the inner portions of Grand Canyon, as well as to dependable sources of water, so it should be of little surprise that some of the earliest mining work was done here.

Bridal Veil Mines

Although prospectors may have been in the Cataract/Havasu Canyon area since the time of the Ives Expedition in 1858, the earliest mining operation of record began in June 1873. Charles Spencer located the Moqui Quartz claim in Cataract Canyon, where he found silver and lead in the Redwall Limestone below Bridal Veil Falls on Havasu Creek. These falls today are called Havasu Falls). At first,

only small amounts of lead and zinc were mined, but the hunt-and-peck operation became more productive in the late 1870s.

The first Havasu claim in the new Yavapai County of Arizona Territory was filed on December 6, 1879, for lead and silver. This was the Beckman claim, made by Daniel W. Mooney, W.C. Beckman, H.J. Young, C.M. Marshal, and Alphonse Humphreys. According to Humphreys's niece Mrs. Helen Humphreys Seargent, her father, Matthew A. Humphreys, was one of the early miners in Cataract Canyon. He had told her many stories about the Havasupai, their beautiful homeland at the bottom of the canyon, and the spectacular waterfalls there. He made regular trips into Cataract Canyon.

Although Alphonse Humphreys, with Mooney, Young, Beckman, E.L. Doheny, and two men known only as Potts and Fowler, had been in Cataract Canyon on several occasions, he failed to find a way down around the highest waterfall—196 feet. On a fateful last trip, Mooney took a rope down into the canyon and, trusting his sailor's experience with ropes and rope climbing, let himself down over the falls. Once he was over the falls, the others in the party lost sight of him and the roar of the water precluded any verbal communication. Soon they felt the rope slacken and, running around to the side of the falls, they saw the rope dangling nearly half way down. Mooney lay on the rocks below. Unable to reach him, all they could do was leave (Seargent, 1959).

When Matthew Humphreys took over the claims in 1883, he blasted out a slanting tunnel past Mooney Falls and made steps inside it. This was easier than it sounds; the rock here is travertine, a relatively soft, porous carbonate material that is deposited by the mineral-laden waters of the stream and mists from the falls. Humphreys then set iron spikes into the cliff below the tunnel for hand- and footholds. The tunnel and spikes are still used today to pass Mooney Falls along the Havasu Trail that leads another nine miles to the Colorado River. When Humphreys reached the bottom of the cliff on the west side of the falls, he found Mooney's remains and buried them in the soft earth nearby. A few years later, a flood partially washed out the grave, so Humphreys reburied Mooney farther downstream.

Alphonse Humphreys gave Bridal Veil Falls its descriptive name long before Mooney's death. Sometime after 1914 a flood changed the shape and structure of the underlying travertine and turned Bridal Veil's curtain of water into a single, narrow fall. It was first called by its current name, Havasu Falls, in a photograph published by the Kolb

Figure 19. The Coconino Silver Mine has been known by many names. Photo by Ben Wittick, ca. 1883–1885; courtesy of Museum of New Mexico, Santa Fe.

brothers in National Geographic (Kolb and Kolb, 1914). The name of Navajo Falls, upstream of Havasu Falls, is also credited to Alphonse Humphreys.

One of the most extensively worked claims in the Cataract Canyon area was the Supai Claim, located on a tributary to Havasu Creek. Part of the claim was located as the Beckman Mine, but it was soon forfeited through non-compliance with mining laws. On February 9, 1881, the claim was recorded again at the request of W.W. Jones and Matthew Humphreys, who referred to it as the Coconino Silver Mine (figures 19–21).

In 1898, George Wharton James visited one of the mines in the Havasu area. He noted that this mine, by then abandoned, was "a fairly rich silver mine." Continuing, he wrote, "In rambling about the tunnels I found a couple of boxes and sacks, in which were sticks of dynamite, candles, etc., the latter gnawed by the rats. On the dump are still to be found good specimens of ore, bearing silver and gold in

paying quantities" (James, 1901:299). James's description matches very closely the Coconino Silver Mine. But it is doubtful that James really saw any gold in the samples because no gold has ever been found or detected from this mine (Karen Wenrich, written communication, 1989).

Virtually every tributary along Cataract/Havasu Creek was explored by prospectors. On February 27, 1881, C. Cohen, W.H. Smith, and N. Ellis filed claims above Mooney Falls. Other claims of various names were made between the Bridal Veil (Havasu) and Mooney Falls in 1890, by D.H. Dillon, S.J. Sullivan, John Reese, A.G. Oliver, and Ney Strickland. Their claims, as well as those by many other prospectors, were often previously claimed mines or prospects that for one reason or another had been abandoned.

Mining spread into the upper part of Carbonate Canyon, where three claims were located in 1883 by several miners. On one of the claims, a mine tunnel is located at the base of a prominent spire of rock, which today is recognized as a breccia pipe (figure 22). This impressive tower of

Figure 20. Unknown miners examine lead-silver ore at the Coconino Silver Mine. The hand-tools are typical of all early Grand Canyon mining operations. Photo by Ben Wittick, ca. 1883–1885; courtesy of Museum of New Mexico, Santa Fe.

brecciated rock weathers out in relief against the sheer red cliff—"A Mining Magnet, the Pillar of Gold at Supai" trumpeted the title of an article in Coconino County's official newspaper, the Weekly News (April 17, 1897). The article described this feature as a "dyke," the older spelling of dike, which in proper geological terminology is an igneous intrusion. However, in many items in contemporary literature, written by nongeologists for a general audience, dike was used to describe any rock or mineral exposure that is in some way visibly distinct from the surrounding rock, regardless of its origin. The newspaper reported that J.C. Brown, M. Page Minor, and C.H. Randsburg had struck rich gold ore in their mine in the Randsburg Mining District. That district apparently only included Carbonate Canyon, and this is another case where a new mining district was named off the cuff, despite the fact that this one was just one mile from all other claims near Havasu Falls (figure 23). The article further stated that "Among other mines being developed are the Honda and Grand site, the latter better known as Sullivan's Chimney, it being a pillar of free milling gold and silver ore in stupendous dimensions. It stands over one thousand feet above the rim of Carbon gulch." In this area, additional claims were made during the summer of 1885 at Beaver Falls on Havasu Creek, about two miles downstream from Mooney Falls.

In 1970, while working as a park ranger in Grand Canyon National Park, Billingsley explored the Sullivan's Chimney mine, a tunnel about two hundred feet long at the base of the "tower" in Carbonate Canyon. On geologic maps made later he named this feature the Carbonate Pipe. In the mine, Billingsley saw limited amounts of copper minerals. The mine walls are encrusted with gypsum and dolomite; the air was dusty and dry. It is doubtful that much, if any, ore was produced here. One of the likely miners of Sullivan's Chimney was M.P. Minor, who scratched his name and the year 1897 on a boulder beside Indian petroglyphs about halfway between the mine and Havasu Falls.

Platinum was rumored to have been found near the Colorado River, and around 1902 the Grand Canyon Gold and Platinum Company was created by investors back East. One of the old mine tunnels from this grandly titled,

Figure 21. Camp just north of Bridal Veil Falls (Havasu Falls) in Carbonate Canyon, typical of field conditions under which prospectors worked. Photo by Ben Wittick, ca. 1883-1885; courtesy of Museum of New Mexico, Santa Fe.

ill-fated venture is two hundred feet above Havasu Creek. For about one hundred feet, the tunnel follows a two-foot-thick layer of green glauconitic sandstone (glauconite is a ferrous aluminum oxide) sandwiched between limestones of the Gateway Canyon and Havasu Members of the Muav Limestone, opening into a small "vestibule" at its end. On the slope below the tunnel entrance is a fan of rock debris reaching to the cliff edge.

There is another tunnel above Havasu Creek near the Colorado River. Together, these tunnels presumably were exploration holes and claim assessment work. The platinum, however, has yet to be found. Billingsley noted much mining equipment in the tunnels in 1970, but in the back of one tunnel there was only trash (some of it modern) when visited by Spamer in the summer of 1989 and subsequent visits.

In the fall of 1982, unidentified backpackers discovered a nineteenth century cache of miner's tools and personal possessions in the lower part of Havasu Canyon. The supplies had been left beneath an overhang along the Havasu Trail on the east side of the canyon three or four stream crossings upstream from the Colorado. A brief handwritten note dated January 5, 1893, indicated that the man was S.C. Rees. John Reese (note the difference in spelling) had claims with several partners farther up Havasu Creek in 1890, and this may have been a relative. The cache contained the necessary mining tools, dynamite, and cooking equipment, in addition to other amenities: a bottle of beer, an anti-drinking pamphlet, matches, candles, playing cards, soap, medicines, buttons, and "girlie" pictures in a remarkable state of preservation (Robert C. Euler, oral presentation to the Arizona Historical Society, Kingman, 1984).

On September 23, 1885, Phillip McDonnel and Cornelius Clearea filed placer claims near a fork of Cataract

Figure 23. *View looking east into Carbonate Canyon from the rim of the Esplanade Sandstone. The Carbonate Pipe is at arrow A, the Coconino Silver Mine is at the bottom of canyon at arrow B, and Havasu Falls is just out of view at arrow C. Photo by George Billingsley, 1970.*

Canyon about thirty-five miles north from Ashfork, but nothing came of these claims. In the upper part of Cataract Canyon, more claims were filed on September 7, 1892, and January 1, 1893, by S.J. Sullivan, Paul Dillon, J.R. Dillon, S. Morrison, William Hull, and Ney Strickland, among others. Again, the claims were wishful thinking.

In the early part of the twentieth century, W.I. Johnson of Prescott was a freighter operating into the American Indian country of northern Arizona. Having heard of high-grade lead deposits in Havasu Canyon, he acquired the claims, organized the Northern Arizona Lead and Zinc Mining Company, and sold stock around Prescott. Johnson, being a man of high ambition, planned to harness the beautiful Bridal Veil Falls for power, and imagined an aerial tramway to Watahomigi Point on the rim. But mining production was meager and unprofitable.

Johnson did make some use of the water rights he claimed, namely, a ditch from Bridal Veil (Havasu) Falls

Figure 22. *Carbonate Pipe—the "Pillar of Gold at Supai," according to the boisterous headline in the Coconino County Weekly News (April 17, 1897). A mine tunnel (arrow) is at the base of this breccia pipe. Note that the "pillar" of rock in the canyon wall, also known as Sullivan's Chimney, is the exposed part of the pipe that is more resistant to erosion, weathering out in relief from the relatively softer sandstone canyon wall. Photo by George Billingsley, 1970.*

from which he operated a small homemade concentrator consisting of a jig operated by an overshot waterwheel. He ran through his concentrator perhaps several hundred tons of lead ore. Some of these concentrates were still on the ground in 1922 when they were found to contain 60 percent lead with perhaps twenty ounces of silver per ton (Busch and Ferris, 1923). Around 1946, the "field office" of the Northern Arizona Lead and Zinc Mining Company was still standing, a single abandoned cabin between Havasu Falls and Mooney Falls, with nothing more than a few scattered papers and the spirits of Mooney and Johnson (Corle, 1946).

In October 1919, Johnson interested C.A. Heberlein of Prescott in his claims. Heberlein, a mining engineer said to have considerable experience with limestone deposits, became convinced that the property offered possibilities for mine development. He recognized, though, that the ore deposits and the extent of the property were too limited to invite the capital needed for the assurance of success in this difficult region. In January 1921, Heberlein took over a definite option to purchase the Johnson claims, organizing the Bridal Veil Mines Association of New York City. The claims purchased were the Bridal Veil, Cataract, Seligman, Pyramid, Iron Mask Lode, Blende, Elania, Home, and Road claims. Until that time, about fifteen thousand dollars had been spent in actual development work. Now, Heberlein also recognized the presence of valuable vanadium ore; tests indicated fifteen to twenty-five pounds of vanadic acid per ton (Busch and Ferris, 1923). Assays of the various claims are listed in Table 2.

In a 1926 letter to a friend, the postmaster at Grand Canyon Village referred to the mines in Supai Canyon (Havasu Canyon) as the Johnson claims (Grand Canyon National Park, Research Library Special Collections): "The mine has been worked but little due to the fact that those who hold it believed it necessary to have a batch of trained eagles to get the ore on top. Johnson has been trying to float stock enough to start operation, but he is one of those fellows who couldn't sell gold out of the U.S. Treasury if he had the opportunity."

Around this time, another interesting Havasu development was unfolding. In 1926, the Fred Harvey Company, the corporate hosteler at Grand Canyon Village, was planning to establish a tourist camp at Supai, the hometown of the Havasupai. They planned to build a road into Cataract Canyon, undertaken by the Grand Canyon Gold and Platinum Company. The road was worked on between 1927 and 1928, but it did not go very far. Starting at the head of Lee Canyon, an eastern tributary to Cataract Canyon, work began in two places a mile apart, one at the top of the four-hundred-foot-high cliff of Coconino Sandstone, and the other at the bottom of the cliff. The route itself is ancient, and the twentieth century work only was improving a new alignment to the old trail. The roads were to meet in the middle, but they never did. The offset is about two hundred feet. About seventy-five thousand dollars was expended on this unsuccessful road, but even though it was a failure, a trail (now called the Topocoba Trail) was sufficiently cleared and rebuilt so that hikers and pack animals can reach Supai today.

Still another early story of the mining paradise in Cataract Canyon took place around 1900. Miners constructed an impressive iron scaffold below Mooney Falls (figure 24), on which they placed wooden ladders. The scaffolding rises 258 feet straight up the Redwall Limestone cliff to the tunnel of the Cataract claim. The thirty-foot-long tunnel was driven into a vanadium ore zone that elsewhere would have been valuable, but here the costs of mining and delivery made the venture unprofitable. A thin path was blasted sixty feet along the sheer face of the cliff to another tunnel (Birdseye, 1925). Busch and Ferris (1923) reported that the vanadium ore had first been found on the sheer cliffs by shooting down chunks of rock with a high-powered rifle, firing at places that seemed to contain ore-bearing minerals. A small side canyon with a spring is located near the iron scaffold. Referred to in 1887 as Mooney Canyon by A. Chewning, this canyon was probably used as a base camp for activities up the ladders.

Mining operations in Cataract Canyon began to decline in the early 1900s. During World War II, a new attempt was made to produce lead and vanadium from the old Havasu claims. Under the auspices of the Havasu Lead and Zinc Mining Company, approximately 156 tons of high-grade oxidized lead ore was packed on Havasupai ponies and hauled up the eleven-mile trail to the rim, from where it was trucked to El Paso, Texas. The operation failed, as usual, because of the high costs of mining and transportation from such a remote locale. Inadequate financial returns, and an engineer's report that indicated that the ore reserves were insufficient, forced the sale of the claims. They were purchased by the National Park Service in 1957, bringing to an end the long history of mining activity in Cataract/Havasu Canyon.

Around 1970, Billingsley was a park ranger stationed in Havasu Campground, in the national park just north of the boundary with the Havasupai Indian Reservation. He explored the old mines and found them unsafe and caving in. Campers had dumped tons of trash into the tunnels and shafts. Lead-silver ore and sphalerite with large dog-toothed calcite crystals were still present in some of the tunnels, but they had been vandalized.

The mining activities near Supai were also the source of much grief to the Havasupai Indians. Land disputes developed between miners, Indian Agents, the U.S. military branches that had cognizance over the area, and the territorial governments (Records of the Adjutant General's Office, Record Groups 94 and 407, Letters Received by the Office of the Adjutant General [Main Series], 1881-1889). The disputes, which involved areas on the adjacent plateaus, too, were not resolved until the twentieth century. In 1975, the mines, canyon, and adjacent plateau were returned to the Havasupai, who do not permit visitation to any of the mines within the boundaries of their reservation.

Hacks Canyon Mining District

Wandering prospectors investigated the area north of the Colorado River and the Havasu Canyon Mining District during the late 1880s. They met the added difficulty of the scarcity of running water, but it was not completely a deterrent. The area is near to the settlements of Pipe Springs and Fredonia, which had the distinct advantage of shorter haulage distances to wagon roads.

Just north of the boundary of the Hacks Canyon mining district as depicted in Map 2, Billingsley has recently discovered a small copper mine about eighteen miles north of Toroweap Canyon, in the White Pockets 7.5 minute quadrangle. The mine consists of two small shafts dug into a sandy limestone bed of the middle Harrisburg Member of the Kaibab Formation. Near one of the shafts, which is about thirty feet deep, is a pile of debris containing copper minerals, mostly malachite and azurite. Judging by the remnants of trash left in the miners' camp, the mine dates from the 1940s. It was claimed for uranium exploration by Energy Fuels Nuclear of Fredonia sometime in the early or mid-1980s. The company dug numerous exploratory trenches near the mine, and drilled

Figure 24. Iron scaffolding is attached to the Redwall Limestone cliff below Mooney Falls on Havasu Creek. The scaffolds once supported wooden ladders that were used to reach a lead, silver, and vanadium mine just out of view at the top of this photo. Most of the wooden ladders were gone by 1970, but several had still been present in the upper half of the scaffolding in 1964. Photo by George Billingsley, 1970.

into a breccia pipe nearby. It is not known whether any uranium minerals were found there, and in 1994 the claim appeared to be abandoned.

The Kanab Gold Rush

Most of the claims in Grand Canyon, as we have seen, were located for copper and, later, uranium, but "Gold!" has always been a rallying cry to prospectors. In 1872, when John Wesley Powell made his second run down the

Colorado River through Grand Canyon, he briefly employed George Riley and John Bonnemort to assist in the surveys he was making of the area. In October 1871, while waiting for Powell's party to arrive at the Crossing of the Fathers (now inundated by Lake Powell), Riley and Bonnemort, out of habit, panned the sands along the Colorado River and found a slight showing of gold. Later, when they reached the mouth of Kanab Creek (Mile 143.5 in Grand Canyon), they panned out "a few colors," but the gold was extremely fine, like flour. Even so, word of the discovery reached the outside world and by February 1872 hundreds of prospectors were arriving at Pipe Springs, Arizona (Cleland and Brooks, 1983), and Kanab, Utah (Beaman, 1874). They poured down Kanab Creek, consumed by gold fever. The rush lasted four months before the truth was realized: gold was there, and flakes could be raised from the creek sand almost anyplace, but the work was extremely hard, and the returns exceedingly small. Disillusioned, the prospectors left, following richer dreams.

The short-lived gold rush was mentioned in a travelogue by E.O. Beaman, who was working with John Wesley Powell's surveyors in the area north of Grand Canyon. Beaman arrived in the midst of confusion and provided the first descriptive account of a journey down Kanab Canyon to the Colorado River (Beaman 1874:590):

> Shortly after leaving Salt Lake a rumor had reached the city that gold had been struck at the mouth of Kanab Cañon, and that miners were taking out two ounces of "dust" daily. Although, from my knowledge of the Colorado cañons, I had little faith in the reported discovery, yet I was prepared to find Kanab in a blaze of excitement. I was not in this disappointed; for upon my arrival the town presented much of the same animated appearance as California did in '48. But a week of prospecting served to cool the excitement, and the majority of the miners left in extreme disgust.

The gold frenzy had sent wishful prospectors into the Colorado River corridor, too. Near Kanab Creek is a miner's cache that was likened to a "movie set" by National Park Service workers in 1990 during the Grand Canyon

River Corridor Survey Project (Fairley et al., 1994). It is "a pristine nineteenth century miner's camp/equipment cache . . . situated under an overhang on a narrow bench overlooking a river" (Fairley et al., 1994:196). Examination of the site in 1994 showed that the aluminum tag left by the 1990 survey crew identifying the site with an Arizona archaeological site number had been chewed nearly through the middle by packrats. Ironically, it had lasted not nearly as long as the old prospecting equipment.

Figure 25. Prospector's cache, probably left over from the Kanab Creek gold rush of 1872. Photo by Earle Spamer, 1994.

The site (figure 25) was apparently first recorded in 1937 by Edwin D. McKee, then the Grand Canyon National Park naturalist, during the Carnegie Institution of Washington- California Institute of Technology expedition down the Colorado River. McKee came upon the cache while measuring a stratigraphic section of the Muav Limestone with Robert P. Sharp (the Sharp diary, November 13, 1937, does not mention the cache; Northern Arizona University Library, Special Collections). A two-page typewritten expedition report by McKee itemized "many rusty implements," including "knives, forks, a pan, two large dippers with improvised wooden handles, a cold chisel, two large sluce [sic] boxes with holes pounded through a curved iron bottom" (McKee memo, December 11, 1937, Grand Canyon National Park; on microfilm at the Technical Information Center, Denver Service Center, National Park Service, original D-225 in storage). The 1990 NPS survey carefully identified the relics and recorded their exact placement at the time. Heavier equipment, such as picks or a dutch oven, have not been found. They probably were taken out by the miners when they quit, although Fairley et al. (1994:196) indicated "The materials present seem to have been left with the intention to be used again upon the owner's return." The NPS survey concluded that the camp had been occupied by two men, based on the duplication of some of the relics. Evidence provided by the artifacts allowed the investigators to estimate that the site had been occupied probably shortly after 1872 (Coder, 1994). Although the operation was principally a placer-gold prospect, small holes and retaining walls nearby suggest that gold deposits also had been looked for, but "If gold had been extracted in anything but the smallest quantity, word would have leaked out" (Coder, 1994:119).

Beaman appears to have met one of the men who operated this placer operation. He wrote that on April 16, 1872,

I visited the mining camp, of which one John Riley was chief…. Expecting to find them hard at work "panning out," we were somewhat surprised to find only one person in camp, Riley having gone up the river a week previous with a small rocker to work up a newly discovered flat, and the others of the company being absent on a "prospecting trip." (Beaman, 1874:591)

The allusion to a larger party may have been bluster, or an attempt to dissuade Beaman from claim jumping. Thereafter, it is of note that Beaman and an assistant traveled eastward to discover Surprise Valley. They visited Deer Creek and its waterfall, first seen by Powell during his 1869 river expedition. Beaman reported that it was known locally as Marble Cañon, which duplicates the name used for the Colorado River gorge in the easternmost part of the national park. Upstream from Deer Creek, Beaman and partner climbed around what is now called the Granite Narrows section of the Colorado River. Had they gone farther, they may have discovered the asbestos deposits along Tapeats Creek and upstream along the Colorado River, which, amazingly, were never prospected.

Hacks Canyon Mine

East and north of the Mount Trumbull Mining District, and north of the Colorado River, copper deposits have been known for a long time in Hacks Canyon (Hack Canyon in some records), a tributary to Kanab Canyon. The mines that were developed here were situated on two separate breccia pipes in the Hermit Shale just above the Esplanade Sandstone. (Later investigations revealed additional pipes in the area.)

According to Aldus F. "Blondie" Jensen of Fredonia (personal communication, 1975), he helped build the road down Hacks Canyon to the mine when he was "a lad" (he was born in 1892), using a team of horses and a wagon. When the road was built, the mine was called the Spotted Bull. Jensen did not know who first located the ore, but he guessed it was discovered around 1890.

During World War I, copper-bearing exposures in Hacks Canyon were relocated. Dunning (1948) took note of the remains of an old tramway that went into the canyon, which may date from this period. About 559 tons of ore were produced from the Hacks Canyon Mine at this time, yielding 51,060 pounds of copper, 1,210 ounces of silver, and two ounces of gold (Chenoweth, 1988:9).

Because copper prices were rising in 1936, Blondie Jensen and his associates, Clair Pierson of Fredonia, Ray Pointer of Klondike, Arizona, and G.C. Harwood of Phoenix decided it was time to develop the claims. Between 1937 and 1946 they acquired the property, which consisted of fourteen patented claims comprising the whole of the Hacks Canyon Mine. The Copper Lady was located on September 24, 1937; the Hacks on March 1, 1941; the Hacks No. 2 on April 12, 1941; the Hacks No. 3 and No. 4, and the Mystery, on September 5, 1941; the Hacks No. 5 on March 16, 1943; and the Copper Lady Nos. 1-5, and the Mystery Nos. 2 and 4 on February 26, 1946 (Collins, 1951).

The prominent green mineral in the old Hacks Canyon Mine is torbernite, a uraniferous copper mineral. The Hacks Canyon uranium was discovered by examining the workings with a "black light" fluorescent lamp in 1945; a sample assayed at 1.56 percent uranium oxide (Dunning, 1948). It was one of the first discoveries of commercial uranium in Arizona outside of the established uranium mines in the Four Corners area. Uranium, when first commercially produced in the 1800s, was used for making a stunning black paint for ceramics, but the brooding secret weapons industry of World War II was about the change the color of things forever. New deposits were examined with much interest.

At first, there was no attempt toward systematic mining at the Hacks Canyon Mine, and it is difficult to understand why some of the work was done at all. In July 1948, Charles H. Dunning, director of the Arizona Department of Mineral Resources, mapped and sampled the mine and

Figure 26. The Hacks Canyon Mine headframe as it appeared in 1948. Photo courtesy of Charles H. Dunning, Arizona Department of Mines and Mineral Resources.

prepared a report (Dunning, 1948) of his investigations for the office of the Atomic Energy Commission at Grand Junction, Colorado. He described the mine (figures 26, 27) as containing a 90-foot vertical shaft, a 55-foot-long adit that connected to the shaft at the 31-foot level in the Hermit Shale, an inclined shaft that reached the 105-foot level, and 400 feet of several winding underground drifts on the 59-, 69-, and 105-foot levels. An inclined winze led to another drift, connecting with yet another drift from an inclined shaft some 125 feet to the north. This maze of workings indicates either that the miners were merely digging out only the richest ore wherever they found it—"high grading" in mining parlance—or that they were lost most of the time.

Under the name of the Canyon Copper Company, the Hacks Canyon Mine produced a few shipments of ore during 1944 and 1945. Between 1951 and 1954, 1,170 tons of copper and uranium ore were shipped by truck to the Atomic Energy Commission ore-buying station at Marysvale, Utah, then by rail to Salt Lake City, Utah, for processing. During this time, however, the road into Hacks Canyon was washed out, temporarily halting production. In 1959, Mrs. Helen J. Bennett of Fredonia, daughter of Blondie Jensen, was secretary of the Hacks Mining

Company, which had been formed on July 13, 1951, in Phoenix. The original stockholders were A.F. Jensen with one-half interest, Elliott Pierson with one-eighth interest, Clair Pierson with one-eighth interest, and G.C. Harwood Trust with one-quarter interest (Collins, 1951). Foord et al. (1978) had indicated that the Canyon Copper Company prospect delivered an unknown amount of copper.

The mine changed hands frequently during the next few years. In December 1952 it was leased to the California Tungsten Corporation, which later merged with the Consolidated Uranium Company of Salt Lake City. The company did some improvement work on the property before forfeiting their lease in the spring of 1953. The Vanadium Corporation of America, of Durango, Colorado, took an option on the property in May 1953, but they dropped it that July. The LaSalle Mining Company of Grand Junction, Colorado, leased the claims on September 7, 1953, and purchased the equipment for $4,200. During its mining operations, LaSalle dug a one-hundred-foot-long, thirty-foot-deep trench east of the old adit (Chenoweth, 1988).

In April 1954, LaSalle subleased the property to Rainbow Uranium of Salt Lake City. That August, Rainbow was reorganized into Urainbow, Inc., which subsequently subleased the property to a group of miners. Much of the uranium ore that Urainbow shipped to Salt Lake City was below the minimum economic grade of .10 percent uranium oxide, making this an unprofitable venture.

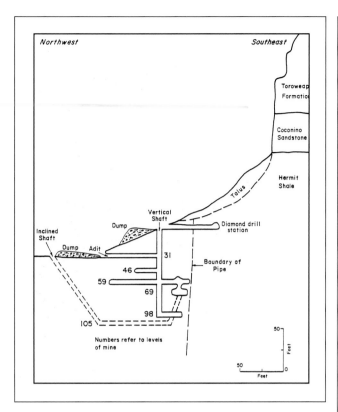

Figure 27. Profile of the Hacks Canyon Mine, showing its levels and relationship to the breccia pipe. From Chenoweth, 1988, figure 4. Modified from C. Rasor (1949) with additions from H.E. Nelson and S.R. Steinhauser (1952) and Ben Bowyer and W.R. Peterson (1959).

Eventually, a lawsuit developed between the Hacks Mining Company, LaSalle, and other parties, in which Hacks Mining gained control of the property and returned the mining equipment as part of the settlement (Chenoweth, 1988).

The Hacks Canyon Mine remained idle for three years after the lawyers and bureaucratic record keepers made most of the profits from the mine. On September 5, 1957, the Rare Metals Corporation of America, of Salt Lake City, obtained a ninety-day lease, which they dropped after a little exploration. Western Gold and Uranium was the next group to have expectations from the mine, leasing it in June 1958. By then, it was known that the mine was in a breccia pipe, and Western Gold already had the experience with pipes from the Orphan Mine near Grand Canyon Village (Grand Canyon Mining District). However, because only low-grade mineralized rock was encountered in the Hacks Canyon Mine, Western Gold dropped the lease early in 1959 (Chenoweth, 1988).

The next lease was to Atkinson Exploration Company of Cedar City, Utah, in the summer of 1959. They did consid-

erable drilling, but no ore was developed and the property was returned to Hacks Mining. After a brief examination, Ivor Adair of Moab, Utah, doing business as the A and B Mining Company, leased the mine from Hacks Mining on September 5, 1963. In 1964, A and B repaired the Hacks Canyon road and in March began shipping ore from the old open cut; 206 tons were sent to Salt Lake City, averaging .13 percent uranium oxide (Chenoweth, 1988). The property remained idle from 1964 to 1972.

In 1971, when Billingsley first visited the mine (figure 27), the road was badly washed out. The tramway reported by Dunning (1948), on the north face of the canyon wall near a trail site, was still present, but unusable and little equipment remained. The ghost mine and many of the mine shacks and construction equipment were rusting or falling apart (figure 28). The mineral torbernite was seen together with spotted clusters of malachite and azurite in the red sandstone and shale. The only water supply for the area was a small spring about three miles farther down Hacks Canyon, where tanks were constructed to water cattle.

On December 1, 1972, the Hacks Canyon Mine was acquired for two million dollars by Western Nuclear of Denver, Colorado. In 1975, the mine was taken over by Energy Fuels Nuclear, also of Denver. Exploratory drilling around the old mine in July 1977 revealed that significant amounts of uranium oxide ore were present in an adjacent breccia pipe. Mining began at this pipe, the Hack 1, in January 1981, and continued until the ore was exhausted in August 1984. Uranium prices gradually dropped in the early 1980s, but the Hacks Canyon Mine contained ore rich enough to sustain operations at a profit in spite of the 320-mile haul to the mill at Blanding, Utah.

Foord et al. (1978) listed four workings for the Hacks Canyon deposits: Hacks Canyon Mine in sec. 7, T. 37 N., R. 4 W., having produced only "Less than 1000 tons" of copper; Hacks Canyon Mine listed only as in T. 37 N., R. 4 W. and 5 W., production amount unknown; Hacks Canyon Mine in NE1/4 sec. 26, T. 37 N., R. 5 W., producing an unknown amount of copper, silver, and uranium; a "prospect" in sec. 20, T. 37 N., R. 4 W., production amount unknown; and another "prospect" in sec. 2, T. 36 N., R. 5 W., production amount unknown.

But the old mine was not dead yet. In 1979, more exploratory drilling in a third pipe, the Hack 2, farther up the canyon from the Hack 1, revealed an even larger deposit of uranium oxide ore. Mining started again in July 1980 and continued until depletion in June 1987. In response to environmental mandates, the land was reclaimed more to its original state in November 1987. In August 1980, a fourth breccia pipe, the Hack 3, was found by exploratory drilling from the Hack 2 Mine. Mining on the Hack 3 started in November 1982 and the pipe was finally mined out by May 1987 (personal communications, Energy Fuels personnel in Kanab, Utah, 1988).

In late 1988, the era of mining in Hacks Canyon ended. Energy Fuels had completed a reclamation project of the Hack 1, 2, and 3 mines in April 1988. The mining venture was done with faster, modern equipment which resulting in just three years of mining before depletion of ore reserves. If the pick and shovel methods of the early 1900s miners had been used, the operation would have continued for about twenty years. This is one mine in Grand Canyon that made a substantial profit, largely due to improved mining methods.

Figure 28. Some of the buildings at the Hacks Canyon Mine that were probably living quarters, built perhaps in the 1940s or 1950s. Photo by George Billingsley, 1971.

Unmined Minerals in the Thunder River Area

Thunder River is a remarkable tributary to another perennial stream, Tapeats Creek, about two miles north of the Colorado River in the east-central part of Grand Canyon. Together with Stone Creek, a tributary to the Colorado two miles upstream from the mouth of Tapeats Creek, this area has not previously been assigned to a mining district because, surprisingly, no mining activity seems to have taken place there, despite evidence of trailing into the Surprise Valley area reported by Dutton (1882). We include it in the Hacks Canyon mining district.

That this area was never prospected is amazing; it contains visible mineral deposits, plentiful clean water, and is no less inaccessible than many other parts of the Canyon. The fact that water was always a problem for early miners would have made this section of the Canyon very attractive. It is one of the most well-watered areas of the Canyon (Huntoon, 1970). It would have been a prospector's dream.

In the mid-1960s, Billingsley found several veins of asbestos up to three inches in width at the mouth of Tapeats Creek. The veins continue upstream along the Colorado for nearly two miles, to Stone Creek. The asbestos occurs in the Bass Limestone along the contact with a black diabase sill. Green chlorite, green garnet, and talc are associated with the asbestos in the two-to-three-foot thick zone of contact metamorphism. The contact can also be traced upstream along Tapeats Creek, and near Thunder River in 1991 Spamer saw asbestos that had been metamorphosed to the "tiger-eye" variety.

Today's Thunder River Trail from the North Rim, though rugged, gives credence to the statement that this area is relatively easily accessible. It is a mystery why prospectors left no trace here. The asbestos veins occur in precisely the same stratigraphic position and mode of occurrence as do the well-known asbestos deposits at Shinumo Creek-Hakatai Canyon developed by William Wallace Bass in the early 1900s, and those in Asbestos Canyon developed by John Hance around the same time.

In the 1960s and 1970s, feral burros, probably descended from the burros that escaped or were set loose by Grand Canyon prospectors, were numerous on the north side of the Colorado River for about seven miles upstream from Tapeats Creek. The burros discovered the value of plentiful, clean water in these tributaries. It is a wonder, then, that their masters had not.

Grand Canyon Mining District

Some of the mining activities in the eastern Grand Canyon were designated as a mining district. The district as we have organized it for this publication did not originally include all of the mining in this area. To simplify matters, and to be consistent with the discussion of areas eastward across the region, we include all activities in the Bass Trail area within the Grand Canyon Mining District.

Bass Mines

The story of mining in Grand Canyon would not be complete without an account of "Captain" William Wallace Bass (figure 29), one of the pivotal pioneers of Grand Canyon history. As told by his son, the late Bill Bass of Wickenburg, Arizona, Captain Bass first visited the Canyon as the result of a story told to him by Emma Lee, a widow of John Doyle Lee of Lees Ferry and elsewhere (Murbarger, 1958):

Emma who was then residing at Ashfork, Arizona, told Dad that during the several years Lee had been in hiding from the law, he had cached three 5-gallon cans filled with gold nuggets in the Canyon. She had a map, she said Lee had made to mark locations of the cans, and gave either this map or a copy of it to Dad, together with some nuggets Lee had found. Soon after he began his treasure hunt, Captain Bass realized that Grand Canyon possessed an intangible worth greater than all the gold that was hidden and established a permanent home at what is now known as Bass Camp.

Sometime between 1883 and 1885, Captain Bass resided in the western part of the eastern Grand Canyon. With the help of two Havasupai Indians, one called Captain Burro (Corle, 1946), Bass built a trail down the upper canyon walls to Mystic Spring, spending his first winter roaming and studying the Canyon with his supply of geology and astronomy books. He and others of his time called this trail the Mystic Spring Trail. Today it is the upper segment of the South Bass Trail (Babbitt and Thybony, 1991). Mystic Spring itself is misplaced on old Grand Canyon maps and is not mentioned on newer editions. In 1972, Billingsley and Jan Jensen located the lost spring in the Esplanade Sandstone. The water seeps down through the sandstone from natural rock pools about thirty feet above the spring, and trickles into a shallow bowl-shaped depression about the size of a large dish pan. The spring is dry most of the year, dependable only in wet weather.

By the spring of 1884, Bass was convinced that Grand Canyon had great potential as a tourist attraction. The Victorian era introduced the means to travel farther, faster, and in greater comfort than had ever been possible. By attending travelogues and lectures, common forms of leisure

activity in those days, the moderately affluent classes learned that they, too, could travel to exotic places. Catering to the growing cult of tourism and his own growing business, in 1898 Bass extended a more crude trail another seven miles from Mystic Spring to the Colorado River, descending what he called Trail Canyon (today it is Bass Canyon). Later he bought an old four-horse stage-coach and established the first passenger service from the railhead at Anita, carrying passengers to his hostelry, Bass Camp, on the Canyon rim overlooking the Grand Scenic Divide. He also obtained a small canvas boat with which he could cross the river to his mines on the north side.

In 1885, seeking to extend the railroad to Bass Camp, Bass had succeeded in luring to his camp the general passenger agent of the Atlantic and Pacific Railroad. But the agent was not abounding in foresight, for upon returning to San Francisco he reported that Bass's proposal was not feasible. "No one," he declared, "would go that far only to see a hole in the ground."

In the summer of 1894, a young lady arrived at Bass Camp on a tourist excursion. Ada Diefendorf had come west from East Worcester, New York, to resettle and to teach music, and wanted to go into Grand Canyon. Romance and blisters blossomed on the trip, and in January 1895 the twenty-five-year-old music teacher married the forty-three-year-old Bass. She spent the next thirty years of her life taking care of and cooking for family and tourists. She became the first white woman to raise a family on the Canyon rim, and she was the first white woman to ride down the rugged Havasu Trail (Topocoba Trail today) to the Havasupai village of Supai.

Among the chores of hosteler, guide, hostess, cook, laundress, seamstress, chambermaid, and frontier wife, Ada Bass raised three daughters and a son. Water was quite scarce, so she periodically loaded soiled laundry onto a burro and made the rugged eleven-mile trip down to the river and camped there overnight to do the wash. A burro named Joe and a dog named Shep would always tag along (Murbarger, 1958).

Shortly after 1890, Captain Bass located several asbestos claims on the north side of the Colorado River, and soon developed a field camp and planted an orchard along Shinumo Creek. It was this camp that would service his cross-canyon tourist trade, for he was the only Grand Canyon pioneer to provide a route from rim to rim (today the South and North Bass Trails; Babbitt and Thybony, 1991). Although it was generally called Shinumo Garden or Shinumo Camp, some of Bass's guests had more descriptive

Figure 29. "Captain" William Wallace Bass and his dog Shep astride a burro at Bass Camp on the South Rim. Photo by H. G. Peabody, taken between 1898 and 1901. Photo courtesy of Grand Canyon National Park Museum Collection, photo no. 833.

layers of the Bass Limestone, the same as at Hance's mines and the Thunder River occurrence. The mineral was formed more than 900 million years ago by a process of contact metamorphism, from chemical reactions taking place when the then-molten diabase sill intruded into the Bass Limestone at great depth in the earth's crust. The hot diabase chemically altered the dolomite and siliceous impurities in the Bass layer, creating asbestos, chlorite, serpentine, and talc.

Billingsley visited the Bass asbestos mine and the camp near it in 1977, and found most of the camp equipment and tools were still there. Today, all of Grand Canyon's asbestos workings are closed to visitation by order of the National Park Service.

When the Atchison, Topeka, and Santa Fe Railroad finally extended the Williams-to-Anita branch line all the way to Grand Canyon Village in 1901, Bass, a competitor of the Santa Fe's tourist interests at the village, in 1902 somehow managed to get the railroad to put in a flag stop about four and one-half miles short of the village. It is shown on the classic 1906 edition of the U.S. Geological Survey's 15 minute Bright Angel quadrangle, surveyed by François Matthes. At Bass Station, passengers traveling to Bass Camp would leave the train and transfer to horses, wagons, or Bass's four-horse stagecoach for the trip to Bass Camp. Departing visitors would have to flag down the train as it passed on its return to Williams.

In 1908, Bass replaced the cable crossing by building a cage ferry (figure 30). The wooden cage, large enough to carry pack animals one at a time, hung from four main cables attached to a pull cable. Then it became possible to begin working the asbestos mines at a profit. Ore was transported by burros, each animal carrying about ninety pounds. Some of Bass's asbestos showed the highest grade of any asbestos mined to that time, and it was shipped to France to be used in the world's first fireproof theater curtains (Hughes, 1967). In June 1931, one specimen of chrysotile asbestos from Bass's mine was included in a prize-winning collection of Grand Canyon minerals at the Northern Arizona State Fair in Prescott (Waesche, 1934). But in the long run, Bass's mines were not very profitable.

names. Henry A. Pilsbry and James H. Ferriss, malacologists who were collecting land and freshwater snails during a cross-canyon scientific expedition in 1906, called it the Thousand Mouse Camp (Spamer and Bogan, 1993:66, 67).

During 1906, Bass constructed a cable crossing over the river at Mile 108. Normally, river crossings were made in a canvas boat, but the spring and summer runoff from the Rocky Mountains brought driftwood logs in such numbers that boat crossings were dangerous (Bill Bass, as published by Maurer, 1983). The cable was improved in 1908. On the north side of the river, Bass and John Waltenberg mined a small quantity of asbestos from a tunnel about 450 feet above the Colorado. Bass had four claims on the upper asbestos vein, and three on the lower vein. The asbestos is almost identical in type and occurrence to the kind already found by John Hance twenty-two miles upriver, and to unmined deposits near Thunder River twenty-three miles downriver. The veins are in contact zones on the upper and lower surfaces of a black diabase sill that intruded into the

Figure 30. John Waltenberg and two tourists cross the Colorado River at Mile 108 in Bass's cable cage. Photo taken around 1918; courtesy of Jack Fuss, Flagstaff, Arizona.

They were restricted by their limited, intermittent occurrence and remote location, and he made better money in the tourist trade. Even so, in 1917 several tons of asbestos were packed out the trail, hauled to the railway siding at Anita, Arizona, and sold to eastern buyers for fifteen dollars per ton. ²

Bass later leased the mine to E.L. Quist, who operated it during the winter of 1920. A new cable crossing had been installed at the mouth of Hakatai Canyon, just upstream from Hakatai Rapid. This cable and the one at the Bass Ferry upstream at Mile 108 was by the early 1960s considered by the National Park Service to be a danger to low-flying aircraft and were cut down. With patience and keen eyes, the cable at Mile 108 can be seen draped across the cliffs and boulders on both banks, and the supporting tripod can be seen profiled high above the north side of the river (verified by Spamer and Richard Quartaroli in 1994). The Bass Ferry cage, broken and rotted, still can be seen lying on the south bank at Mile 108.

Jack Fuss, who had worked for Bass, was interviewed in 1979 at his home in Flagstaff. He provided the useful historical insight that only oral histories can. In 1917, Fuss had come to Flagstaff from Philadelphia, along the way

meeting and befriending George Blair, a Cherokee. In Flagstaff, they intended to make a trapping expedition to Utah, but they were told that a canyon was in the way. Planning accordingly, they bought three horses and supplies, but during their three-day trip to the South Rim they lost a horse. At the rim of the Canyon they stopped. Fuss said it scared them to death. They found a camp a little farther along the rim, but no one was around. That night, Blair decided to head back home, leaving Fuss alone and puzzled. The next day, Captain Bass showed up and found him there admiring the Canyon. Fuss took up an offer from Bass to work the mines in the bottom of the Canyon, where he met Chikapanagi, a Havasupai. They became great friends, and for a year and a half they mined asbestos in what they called Bullion Canyon. Because of the scarcity of water near the mine, the main camp (figure 31) was set up on Shinumo Creek, a perennial stream. They stayed at their rude field camp only when mining was in progress.

At first, the asbestos was hauled to the cable at the mouth of Hakatai Canyon, and dumped into a bucket which was pulled across the river. Bass, on the south side, would put the rocks into sacks and hang them over the backs of burros. This method was time consuming and laborious and eventually was abandoned. Thereafter the asbestos was packed—three canvas cement sacks on each of twenty-seven burros—to Shinumo Creek. Then one at a time the burros were then pulled across the river in the Bass Ferry cage. The lead burro was always one named Jaco. Fuss had trouble at times keeping his charges from running off with the feral burros that would wander by from time to time. "Even though we always hobbled them," he said, "they would really hop along when the wild burros came around." The cable was still in place in 1971. Around 1972 it was cut down by Wayne Learn of Kingman because it posed a threat to aircraft in the river corridor.

Jack Fuss ended his mining career with Bass late in 1918 when he nearly starved to death waiting for Bass to come down with supplies. He ate cactus for three days. Finally, Jack hiked out toward the South Rim and met Bass coming down with the late supplies. Bass tried to talk him into staying, but Jack had made up his mind; he had had enough. Jack spent many years in and around Flagstaff working as a game warden and a painter. He married Marie in 1922, and they had two daughters. Jack said, "Captain Bass never did pay me for all that work, but it didn't matter; I liked the adventure."

While Bass was prospecting a side canyon west of the Mystic Spring Trail south of the Colorado, he and Chikapanagi discovered a fair showing of copper ore in a well-defined fissure vein in the Vishnu Schist. Later Bass opened the vein, finding in it a good grade of copper sulfide ore minerals, such as bornite, chalcopyrite, and chalcocite. He also found some galena (a lead mineral), silver, and cuprite (copper) in small quantities. Around 1908, Bass packed out about twenty-five tons of this ore on burros. Billingsley in 1968 examined this mine, noting that the tunnel runs southwestward more than one hundred feet along a vertical vein, one to two feet wide, on the east side of the canyon. A fifty-foot vertical shaft was sunk about twenty-five feet inside the tunnel. A second vein crosses the creek bed a few hundred feet south of the first vein. A seventy-five-foot-long tunnel, known as the Hakatai tunnel, was driven in around two hundred feet above the creek bed (Noble, 1914).

During his thirty-five years at Grand Canyon, Bass constructed more than fifty miles of trails in the Canyon, including the first cross-canyon trail. Bass named the northern portion White Trail, after a prospector. During the 1930s and 1940s it was known as the Shinumo Trail, and today it is the North Bass Trail. White Creek, which is the lower part of Muav Creek, a tributary to Shinumo Creek, bears the prospector's name. Bass's trail also led him part way to claims in Shinumo Canyon between White Creek and Flint Creek (Shinumo Creek). The trail also serviced Bass's copper claims below Swamp Point, under the Kaibab Plateau in the Coconino Sandstone and Hermit Shale. (There is much more information about the Bass Trails in a report by Mike Anderson held in the Study Collections at Grand Canyon.)

Figure 31. Jack Fuss at Bass's camp on Shinumo Creek, around 1918. Photo courtesy of Margaret Learn, Flagstaff, Arizona; from the Jack and Marie Fuss Collection.

Just east of the North Bass Trail is a good spring and an old shelter, apparently one of Bass's workings (Dan Cassidy, personal communication to Spamer, 1995). The "mine," however, is little more than a copper staining at the base of the Coconino Sandstone, where the spring is located. The copper appears to be just redistributed minerals, deposited there by ground water that percolated through fractures in the Kaibab Plateau. When Billingsley examined the site around 1970, he noticed no evidence for a breccia pipe there, but at the time their structure and surface expression were not as well understood as they are now.

Unlike many prospector-hostelers of Grand Canyon Bass never gave up on his mines. He waited for better ways to get his ores out of the Canyon. Kolb (1914:230) recorded that "fractures exposed mineral seams and deposits of copper and asbestos on both sides of the river, some of which Bass had opened up and located, waiting for the day when there would be better transportation facilities than his burros afforded." That day never arrived.

When the National Park Service awarded the Grand Canyon lodging franchise to the Fred Harvey Company, the twelve-room Bass Hotel at Bass Camp was ordered removed by the park service, but it was not torn down until 1937. On September 15, 1923, Captain and Mrs. Bass entertained their last paying guests. Soon afterward they moved to Wickenburg. Early in 1926, the Santa Fe Land Development Company purchased Bass's trails, mining claims, mill site, and other properties in the Canyon for twenty-five thousand dollars. Captain Bass died in 1933 at eighty-four. He wished to be buried atop a promontory in the Canyon known previously as Bass Tomb, now Holy Grail Temple. His wish was carried out by his son, Bill, who dropped a copper box containing his father's ashes onto the promontory from an airplane (Murbarger, 1958). Mrs. Bass survived her husband by eighteen years, and she rests now in the cemetery at Grand Canyon Village. A memorial plaque to Captain Bass is to the right of her grave.

Figure 32. Louis Boucher and his mule near Hermit Camp, in Hermit Canyon. Photo by Edward W. Murphy of New York, date unknown; courtesy of Grand Canyon National Park Museum Collection, photo no. 523.

Boucher Mine

Many prospectors who roamed Grand Canyon enjoyed the serenity and the spectacular scenery. These men were rarely seen and seldom heard from unless they settled near a small community. Some were content just to be there. Louis Boucher was one of these hermits (figure 32). Hermit Basin, Hermit Canyon, and Hermits Rest are named after him, as are the Hermit Shale and the Hermit Fault.

Boucher was French-Canadian, born in Sherbrooke, Quebec (Hughes, 1978). In 1891, he settled at Dripping Springs, about a thousand feet below the South Rim (figure 33). Much of what is known about Boucher (e.g., Hughes, 1978) was obtained from a visitor to Grand Canyon in 1932. Edward Murphy, of New York City, spent several months with him in 1890, together with a cowpuncher named Johnny Stevens and a man from New York "who owned a box camera" and who photographed Boucher (figure 32). Murphy and Stevens helped build the trail to Dripping Springs (McKee, 1933).

Figure 33. Louis Boucher's camp at Dripping Springs, west of the Hermit Basin, at the base of the Coconino Sandstone, as seen in July, 1907. Photo by Herbert W. Gleason; courtesy of Herbert Wendell Gleason Collection.

Boucher established the first camp for travelers inside the eastern part of the Canyon, on the Tonto Plateau near Hermit Creek. He built a trail (now called the Boucher Trail) out and around Columbus Point, then down into Long Canyon (now called Boucher Canyon), where he planted an orchard and garden. He also developed a small copper mine (the Boucher Mine) in the Vishnu Schist near his cabin, but he did little work there. Boucher's other claims included a deposit of graphite (location unknown), but none were profitable. He left the Canyon after 1909.

In 1972, Billingsley visited the Boucher Mine and observed a few low stone walls and the remnants of Boucher's cabin just south of a small mine adit near the creek—all that remains of Boucher's presence. The adit on the north side of the drainage showed small amounts of the copper mineral malachite in a shear zone in the Vishnu Schist.

Orphan Lode Mine

The most successful mine is the Nelson limestone quarry, in the Music Mountain Mining District. The second most successful mining operation in the Grand Canyon region was either the Hacks Canyon mines (Hacks Canyon Mining District) or the Orphan Lode Mine. The Orphan Lode has the richer history by far. Perched on the South Rim just west

of Grand Canyon Village, this uranium and copper mine is probably the most-studied of the Grand Canyon mines.

The ore body of the Orphan Lode Mine is located in a breccia pipe that extends vertically to a depth of about two thousand feet (figure 34). More than sixty different minerals are known from the mine, most of them in small quantities. Metals contained in these ores are uranium, copper, lead, silver, vanadium, and zinc (Kofford, 1969). When the mine closed in 1969, it had produced 495,107 tons of ore, including 4,257,571 pounds of uranium oxide averaging .43 percent, 6,680,000 pounds of copper, 107,000 ounces of silver, and 3,283 pounds of vanadium oxide. The value of the uranium alone has been estimated at forty million dollars (Chenoweth, 1986).

Daniel "John" L. Hogan found the Orphan Lode. Born in Syracuse, New York, in 1866, he trekked westward when he was thirteen years old. By the time he reached Flagstaff in 1890, he was an experienced ranch hand. His knee had been shattered in a gun fight in Texas and was repaired with a silver plate. Shortly after arriving in northern Arizona, he and his partner, Henry Ward, became interested in prospecting the depths of Grand Canyon. In 1891 they located an outcrop of copper minerals in the canyon wall 1,100 feet below the rim at the base of the Coconino Sandstone, between what today are Powell and Maricopa Points, two miles west of Grand Canyon Village. On February 8, 1893, they filed a claim, calling it the Orphan, ostensibly because Hogan was an orphan, but Hogan variously referred to it as the Orphan Lode and the Little Orphan Girl (Dodge and McKleveen, 1970). Hogan and Ward built a crude trail consisting of ropes, ladders, and rock steps from the rim and did a little bit of mining in the Orphan Lode claim

A report in the Williams News for July 24, 1897, provides some indication as to the condition—real and remembered—of the trail to the Orphan Lode Mine:

A Grand Canyon Trail That Thrills With Adventure

Upon a recent trip to the Grand Canyon, D.J. Daze, C.L. Porter and E.L. Elmendorf . . . made the descent over Hogan's Slide [the steep slope from the rim to the original adit]. This is the trail over which five women were guided last summer. They are all living yet, but that date will mark mile posts of thrilling memory in their lives. The descent of Hogan's Slide is perilous at any time, for even an experienced and hardy mountaineer. The ladies last summer were led down through a mistaken impression that it was the Bright Angel trail—they did not go down that way from choice.

Mr. Daze says there are vivid remembrances of last year's expedition clinging to the precipices and blades of grass that form a foothold between heaven and eternity. Here and there are shreds of bloomers, a scalp lock or two, an occasional Della Fox curl and other evidences of disaster, left there by people who approached the depths by leaps and bounds.

Hogan's Slide is named after Miner Hogan, who constructed the trail to a mine in the Canyon. How he ever concluded the job is beyond the ken of man.

During later activities, an aerial tram was built down this slope.

When Hogan and Ward sought to patent their claim in 1905, the patent was granted by Theodore Roosevelt, under whom Hogan had served as a Rough Rider in the Spanish-American War (Mineral Survey no. 2004). It contained

20.64 acres, of which about four acres were on the Canyon rim (Chenoweth, 1986). Patent no. 43508, dated March 23, 1906, was granted to Hogan and co-owner Charles J. Babbitt (Brown, 1955), the latter to whom Ward sold his share. Hogan acquired full interest in the Orphan Lode Mine on April 18, 1912, through a quit-claim for Babbitt's interest (Brown, 1955).

Hogan intermittently mined small amounts of copper ore, but after several years he gave up, probably because of his injured knee. Increasing tourism at the Canyon made Hogan realize the economic potential in the scenic splendor, and he constructed for tourists the Grand Canyon Trading Post, later called Kachina Lodge, and Grand Canyon Inn by the subsequent owner.

On August 1, 1946, Mrs. Bertha Madeleine Jacobs of Prescott bought the Orphan Lode from Hogan for fifty-five

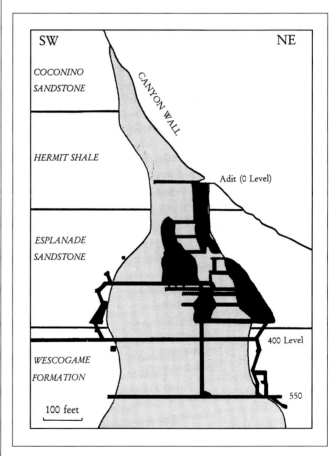

Figure 34. Profile of the Orphan Lode Mine, showing its levels and relationship to the breccia pipe and canyon wall. The shaded area depicts the breccia pipe. Shafts and adits are depicted as seen on a section N 35° E; there also is a service winze between levels 0 and 245, not shown. Large, irregular black areas are stopes, or "clear-cut" areas from which ore has been removed and through which ore is dumped. Level 400 extends to the southwest and the main shaft to the surface. After Chenoweth, 1986, figure 5.

thousand dollars (Brown, 1955). Jacobs was not interested in mining, but envisioned making money from the tourist trade, so canyon-rim property could be valuable. That changed, however, when in 1951 amateur prospectors discovered anomalous radioactivity on the rim west of Grand Canyon Village—just east of where Hogan stockpiled his ore (Kofford, 1969). On April 22, 1951, during the increasingly chilly Cold War, Harry C. Granger of the U.S. Geological Survey examined the property and confirmed the presence of uranium at the lower portal of the Orphan Lode (Chenoweth, 1986). The next decade would see the Orphan Lode transformed for a while into one of the most productive mines in Grand Canyon.

In September 1953, Arthur R. Still, a geologist at Prescott representing the Golden Crown Mining Company (a subsidiary of Western Gold and Uranium of New York), leased the mine and hotel from Jacobs. From Hogan's old adit Golden Crown drilled samples and on December 14, 1955, a letter was sent to its stockholders announcing the discovery of uranium ore in the Orphan Lode. In 1956, Golden Crown acquired complete ownership of the claim, including the Grand Canyon Inn (Chenoweth, 1986), and installed an aerial tramway—each of its buckets could hold eight hundred pounds of ore (Brundy, 1977)—between the rim and the portals at a cost of one hundred thousand dollars.

The first ore shipment, on April 25, 1956, contained 20.89 tons averaging .53 percent uranium oxide, consigned to the Atomic Energy Commission ore-buying station at Tuba City, Arizona, ninety-two miles away. In January 1957, the tramway lifted ore in excess of 726 tons, with an average uranium oxide content of 2.29 percent, the highest monthly grade ever obtained from the Orphan Lode (Chenoweth, 1986). (C.M. Brundy, 1977, has noted that the Orphan Lode at one time yielded the highest-grade single shipments of uranium oxide ore ever in the United States; an average of 4.09 percent, four times richer than other U.S. sources.) The National Park Service also got a piece of the action, charging Golden Crown a fee of two cents per ton-mile for the use of the roads in the national park (Chenoweth, 1986).

Wealth was now coming from the sales of Orphan Lode ore. In July 1957 Western Gold and Uranium acquired full control of Golden Crown. By 1959, a new vertical shaft begun in 1957, was completed to a depth of 1,590 feet, at a cost of about eight hundred thousand dollars (Mining World, 1959). Production levels soared to about 8,999 tons per month instead of the 1,000 tons per month carried by the aerial tram.

In September 1961, Western Gold and Uranium was acquired by Western Equities, Inc. On December 22, 1962, the mine was idled for eleven months when the two-hundred-ton ore bin atop the new shaft collapsed. The shut-down forced the mill in Tuba City to close.

At that time, Western Equities sought permission of the federal government to extend the mine adits beyond the claim boundaries into federally owned land. (In 1958, it had been discovered that the slanting ore veins passed beyond the claim.) To speed the approval process, the company proposed to build a multistory hotel complex hanging over the rim of the Canyon, and Senate Bill S. 383 was quickly introduced in the U.S. Congress. The bill stated that the federal government would acquire the mine in twenty-five years, during which time Western Equities could mine under government land. The bill was tabled, but in February 1960 Arizona Senators Hayden and Goldwater introduced Senate Bill S. 3094, with essentially the same provisions, but it died from lack of action in that session of Congress (Chenoweth, 1986). Bill S. 383 was reintroduced in January 1961, and on May 16, 1962, Congress passed the legislation by a simple majority. President John F. Kennedy signed bill 87-457 into law on May 28, 1962. It would bring ownership of the mine to the National Park Service in May 1987. Mining at the Orphan Lode began again in November 1962, and the Tuba City mill reopened in April 1963.

During the fall of 1963, an incline was driven from the 550-foot level to the 585-foot level, the lowest level to be mined. During 1963, the mine produced 614,858 pounds of uranium oxide ore. Production dropped in 1964, but in 1965 it jumped to an all-time yearly high of 680,746 pounds (Chenoweth, 1986).

Ownership of the mine passed from Western Equities to Geo Space Corporation in 1966 when it became part of a merged corporation, the Westec Corporation. On August 25, 1966, Westec declared bankruptcy and the mine was closed. In August 1967, Westec sold the mine and the employees' trailer park to Cotter Corporation of Roswell, New Mexico, for $875,000 in cash and a 4.2 percent royalty on ore produced. Cotter Corporation reopened the mine in September 1967 and began shipping ore to the mill at Canon City, Colorado, 750 miles by rail at a cost of $12.96 per ton (Chenoweth, 1986). The last ore shipment was made on April 25, 1969, partly as result of spiraling fuel costs.

Records of the Coconino County Assessor's Office show that on February 18, 1981, John R. Siebold of Grand Canyon, Arizona, and Elling Halvorson of Redmond, Washington, acquired the Orphan Lode claim, reportedly paying the Cotter Corporation eight hundred thousand dollars (Chenoweth, 1986).

On May 28, 1987, ownership of Dan Hogan's little copper mine reverted to the National Park Service. Some have suggested that the park service should operate the mine as a historical attraction, but the mine is in serious disrepair and has been sealed. The headframe and some structures are still in place on the Canyon rim, and the immediate area is fenced off with signs warning of physical hazards and radiation dangers. The mine can be seen from the higher vantage of Powell Point to the west, where the park service has installed an interpretive sign.

Bright Angel Trail Claims

The Grand Canyon mining story would not be complete without a few words about Ralph Cameron, "the man who owned the Grand Canyon." Ralph Henry Cameron headed west from Boston in 1883 when he was not quite twenty years old, intent on making a fortune. He and his brother, Niles, arrived in Flagstaff on the new Atlantic and Pacific Railroad and took up operating a sheep ranch and mercantile business. During the spring of 1890, with Pete Berry and a number of other prospectors, they filed numerous mining claims within Grand Canyon, most of which produced nothing.

In 1890 and 1891, Cameron, Pete Berry, and others improved an old Havasupai trail and called it the Cameron Trail (now the Bright Angel Trail). Berry obtained a franchise to operate the trail as a "toll road" in 1891. Cameron acquired the franchise from friends in 1907. Between 1897 and 1903, Cameron built a small hotel on a nearby claim site (figure 35). The Atchison, Topeka, and Santa Fe Railroad completed the spur line from Williams to Grand Canyon Village in 1901, and by 1905 completed El Tovar Hotel on the rim. Rivalry between Cameron and the railroad grew intense.

Before long, Cameron's hotel business floundered, but he believed that his trail could open the way to a fortune. He filed increasingly more mining claims in strategic locations along the trail, thirty-nine of them in a single year, eventually holding claim to thousands of acres, even though there was hardly a mine or placer operation in the area. He controlled the head of the trail, Indian Garden (the only water source along the trail), and the section along the Colorado River. But before he could capitalize on the claims, the railroad took legal action, contesting the claims in a long series of court battles. In 1906, Berry's franchise to operate the trail as a toll road expired, and control of the trail reverted to Coconino County.

By 1912 Cameron believed that his claims would at last pay off. A Philadelphia syndicate took over seven of his

Figure 35. The Cameron Hotel on the South Rim, as it appeared around 1910 (photographer unknown). Photo courtesy of Grand Canyon National Park Museum Collection, photo no. 8155.

placer claims near Indian Garden. Twenty-eight more claims located in 1907 below the Devil's Corkscrew portion of the trail were bonded to New York investors who planned to extract gold. A mine tunnel can still be seen where the Devil's Corkscrew reached Pipe Creek. The National Park Service rebuilt the Bright Angel Trail to bypass the very steep Devil's Corkscrew of Cameron's claims. It enters Pipe Creek farther upstream to descend a little less steeply.

The Santa Fe Land Company finally acquired the key Cameron claims in 1916 from John Daniel, who was acting for Cameron and a partner, C. Frank Doebler of New York. The Santa Fe agreed to pay forty thousand dollars cash to eliminate Cameron's obstruction of their development plans. After Grand Canyon National Park was established in February 1919 Cameron became a U.S. Senator for Arizona. In the following years he continued to attack anyone connected with the National Park Service. In the meantime, Cameron's men still controlled Indian Garden and the Cape Horn claim at the top of the trail, discouraging development. In 1923, suits were filed against Cameron in the U.S. District Court in Arizona by several businessmen and government officials. Two of the suits were contempt proceedings for failure to remove structures and employees from Indian Garden and the Cape Horn claim. When park rangers finally descended on Indian Garden, Cameron's caretakers rushed away, according to stories, leaving six or eight gallons of mash ready for the still.

Carl Hayden unseated the incumbent Cameron in the 1926 Congressional election. Without political influence, Cameron lost all interest in his mining claims. After thirty-five years of fighting to retain control to one of the most economically lucrative parts of Grand Canyon, Cameron quietly gave up. In 1928, the title to the Bright Angel Trail passed to the National Park Service. Cameron died on February 12, 1953, and is buried in the cemetery at Grand Canyon.

An interesting note must be added here about gold near Bright Angel Trail, which was completely missed by Cameron. Waesche (1934) reported that he had managed to pan gold from an ancient gravel bar between Pipe and Horn Creeks, just to the west of Cameron's claims along the Colorado River about one hundred feet above the Colorado's high-water level. "The panning effort netted enough gold and black sand, the sand and the gold being about equally divided, to fill a very small perfume bottle. A prospector would say, 'Yes, there is placer gold along the Colorado but there is too much sand mixed with it to make it worth recovering' (Waesche, 1934:234).

Grandview and Last Chance Mines

Pete Berry came to Grand Canyon by 1888. He traveled the Canyon on foot and on horseback. On his travels he met several prospectors and within a short time he acquired the mining fever. On April 19, 1890, Berry located the Grand View (later Grandview) and Last Chance claims on Horseshoe Mesa, below Grandview Point on the South Rim (cover photo). In addition to Berry, those interested in patenting and locating the mines were Ralph and Niles Cameron, E.I. "Ed" Gale, Robert A. Ferguson, and Thomas McMillan. McMillan, a rancher, put up the grub stakes and was later paid for his share in the mine. Ferguson was soon ruled out because he refused to do his share of the work. In 1892 and 1893, Berry and others built the Grand View Trail (now the Grandview Trail) that led to the mine, along which tourists were later brought down from the Grand View Hotel.

Initial assay reports of the Grandview Mine showed 37 percent copper. At the 1893 Columbian Exposition in Chicago, cuprite from the mine was awarded top prize in a competition for assaying out at more than 70 percent pure copper. The most abundant minerals present at the mine are sulfates such as cyanotrichite, brochantite, chalcoalumite, langite, barite, devillite, chalcanthite, antlerite, and gypsum. Carbonates are also present, notably aurichalcite, azurite, malachite, and smithsonite. Small amounts of pyrite and cuprite are also found. These deposits are now known to be within a breccia pipe in the Redwall Limestone on the Grandview monocline, a flexure in the earth's crust.

In 1895, Pete Berry and Ralph and Niles Cameron formed the Grand Canyon Copper Company. On August 21, 1901, the property—mines, mill site, trail, and Grand View Hotel—was sold to the Canyon Copper Company, a consortium of capitalists from New England and Flagstaff. (However, see also Verkamp, 1940, 1993, who provides a somewhat different chronology.) John H. Page of Phoenix, a prominent land attorney and one-time Territorial Auditor, was president of the company, and Harry Smith was the general manager of the mine and hotel. Page made more than one hundred trips in his youth, driving pack trains up and down the Grandview Trail (Dunning and Peplow, 1959).

Ore from the Grandview Mine was brought up the trail on strings of eight to ten mules or burros, each animal carrying about two hundred pounds and making a trip and a half per day. Pete Berry estimated the cost of building and maintaining the trail at about twelve thousand dollars.

Ore was stockpiled on the rim and later transported to the railroad fourteen miles to the west, then 615 miles by

Figure 36. Mining equipment at the entrance of the Last Chance Mine. Photo by George Billingsley, 1969.

rail to El Paso, Texas. The company speculated on the construction of an aerial tramway or telpherage system, with electric power generated from Cottonwood Creek, which flows through one of the nearby side canyons to the Colorado River. The major concern was whether they could build a dam that could withstand the large pressures of flash floods (Day, 1905).

But the point was moot, for in 1907 copper prices dropped from twenty-four and one-half cents to twelve and one-half cents per pound, detrimental to the financial success of the mines, and work at the mine ceased. Records show that the Grandview Mine produced about seventy-five thousand dollars in copper. Activities at the Grandview Mine ended in 1916.

The properties were sold to newspaper magnate William Randolph Hearst in 1913. Berry and his wife, Martha, stayed on as caretakers. In 1919 they moved to their son Ralph's ranch and remained there until their deaths, Martha in 1931 and Pete in 1932. They, and Ralph who died of influenza in 1919, are buried in the Grand Canyon cemetery.

After considerable legal wrangling, the National Park Service bought the properties from Hearst in 1939 for eighty-five thousand dollars. The park service removed the rim buildings in 1959 and left the Horseshoe Mesa site as it was.

In 1951, Dr. Russell Gibson of Harvard University contracted with the Atomic Energy Commission to make a radiometric reconnaissance and evaluate the uranium content of "red bed" copper deposits in the southwestern

United States. He examined thirty-six properties in four states: Arizona, New Mexico, Texas, and Oklahoma. The old Grandview copper mine was the only copper mine or prospect in the Grand Canyon region studied by Dr. Gibson, and it exhibited the greatest uranium concentration of all the properties examined (Gibson, 1952).

Billingsley first visited the Grandview Mine in 1965 and found that one cabin, made mostly of stone and wood with a metal roof, was still standing on the narrow neck of Horseshoe Mesa. The roof was about to fall in, which it finally did in the late 1960s. East of the main cabin were the remains of several tent cabins, all flattened and rotting. A mine cart on narrow rails and a metal wheelbarrow were found inside the entrance to the mine tunnel. Numerous shafts and tunnels branch off the main tunnel where considerable stoping had been done. The rails ran back more than three hundred feet to a vertical shaft that reaches the mesa surface about sixty feet up and not far from the cabin. Picks, shovels, and drills were in the mine tunnel. The lower tunnel had several rusting pieces of machinery at its entrance, along with some mining tools (figure 36). This tunnel goes back at least 180 feet toward the southwest, where it ends abruptly. There is a crosscut tunnel going east a short distance, and one vertical shaft inside the main tunnel. Very little copper ore was seen in this tunnel, but some was found in fissure fillings and cracks in the limestone.

By September 1990 when Spamer visited Horseshoe Mesa, the minor workings on the neck of the mesa were in shambles, slumped and blocked by debris. One of the clogged openings could have been the vertical shaft to the Grandview Mine main tunnel. Only the roofless stone cabin remained clearly seen, and a couple of dumps of miner's trash lay scattered. Small pieces of malachite and azurite were strewn about, a poorly marked trail of crumbs leading to the main mine.

The Grandview Trail continues down off the east side of Horseshoe Mesa, across a small side canyon to Miner's Spring, a seep dug out of the Bright Angel Shale under a ledge of Muav Limestone. Farther down the trail, there is a panel with early visitors' names etched in mud-coated walls of Tapeats Sandstone.

Prospectors had several camps scattered throughout this area of Grand Canyon, on both sides of the Colorado River. One such camp was found by Harvey Butchart on top of the Redwall Limestone west of the Grandview Mine. There are very likely other camps in the area, not far from a prospect or from water.

Hance Asbestos Mines

Unquestionably the most colorful character to dominate early Grand Canyon history was "Captain" John Hance (figure 37) who, according to some stories, came to the Canyon from Texas around 1882. Other information puts Hance in the Verde Valley of central Arizona in 1868, and that indicates he drifted up to the Canyon after 1876, but certainly was there by 1883 (Mike Anderson, written communication, 1995). He is the subject of anecdotes and apocryphal tales and was literally a legend in his own time (Lockwood, 1942).

The thing Hance enjoyed most was storytelling. To say he did this well is an understatement. Many stories are ascribed to him. Some may have been his, others given to his memory by admirers. He had stories about his faithful horse, under a variety of names, which could gallop full speed to the rim, leap, and successfully land on the other side of the Canyon. At other times, Hance would recall, "I wasn't half way across before I saw that he couldn't make it. We hadn't taken a big enough start, so I turned him around and went back." In other renditions, Hance and faithful steed fell into the Canyon, but Hance saved himself by dismounting just before reaching the bottom! Anything was possible. After all, Hance, by his own admission, also had dug out the Canyon—but that's another story (Lockwood, 1942).

John Hance began exploring and prospecting in the eastern part of Grand Canyon between 1883 and 1890. During this time he befriended several other prospectors, and together they located the first of several claims in Grapevine Canyon, south of the Colorado River.

Prospectors William H. Ashurst and J. Marshall made several claims with Hance in the Red Canyon area during March 1891, and later claimed most of the tributary canyons on the north side of the Colorado. C.H. McClure, M.M. Fisher, and T.C. Frier joined in several of the claims. Ashurst is known to have actively worked his claims near the Colorado River, but he failed to find any economical deposits. He was killed by a landslide near his mine on January 18, 1901 (Coconino Sun, February 23, 1901; Way, 1980); his decomposed remains were buried "about thirty feet from the river" (Weekly News, February 23, 1901).

On January 1, 1894, John Marshall, C.H. McClure, T.C. Frier, John Hance, and W.H. Ashurst located and claimed a toll road or trail twenty-seven miles long from the South Rim to the bottom of the Canyon, indicating it to be in the Grand Canyon Mining District. The starting point was about three-quarters of a mile east of what is now called the Hance Trail or Red Canyon Trail. (The original "Old Hance Trail" was partly obliterated by landslides in the early 1890s and was rerouted by Hance. The newer trail (Hance Trail) survives, but it is not maintained and is very rugged. The old trail is now barely even a route.) A crudely drawn map was submitted with the claim record to the Coconino County Recorder's Office in Flagstaff. This map, redrawn for clarity by Dave Pemberton for Menkes, shows the toll trail descending Red Canyon, crossing the Colorado River near the bottom of Hance Rapid (Mile 77), and continuing westward on the north side to Clear Creek. In 1982, Menkes found portions of the trail to be still evident between Asbestos Canyon and Clear Creek.

Hance had claimed some asbestos workings on the north side of the Colorado River, in geological occurrences identical to the asbestos deposits mined by William Wallace Bass twenty-two miles to the west and the remarkably unmined deposits near Thunder River forty-five miles to the west (see earlier herein). But he had a problem ferrying his asbestos across the Colorado in a small canvas-and-wood boat. A rendition of Hance's first river crossing is recounted by Godfrey Sykes (Lockwood, 1942):

John Marshall, I think, was one of the original locators of the big ledge (asbestos mine) below the rapid. I forgot what they called this ledge, but at any rate you could see it across on the other side. Once Bill Ashurst made locations on it. The idea was to try to get across. The prospectors wanted to make a boat, so they tore up the floor of the old Hull ranch house for this purpose. They took the boards out to the canyon to make this boat, but it was not a success. No one would get into it. I now came along with the frame of a canvas boat I had made. One after the other, we went across to make locations on the ledge. On one of these trips I took John over. In midstream, he told me that it was the first time he was ever across, yet in his stories he always asserted that he had valuable mines over there.

This may have been the inspiration for the story of an immense copper ledge and an astonishing gold mine, which we recount in this publication in the section on Lost Mines and Hidden Treasures.

working to the rapids, though, is just two hundred feet or so above the river. It consists of a small hole in the cliff along a small bench, about big enough for two people. When Spamer visited the site in 1989, a wheelbarrow and a hand-operated pump manufactured in New Jersey were still in the hole. Tools and asbestos fibers more than an inch long lay scattered nearby.

Most of Hance's development work was near the western end of the nine-thousand-foot-long property, on what is known as claim 14, or Wool Claim, about eight hundred feet above the Colorado River. Billingsley visited the mine workings and camp in January 1969. The camp is located not far below a good spring. He saw several tent foundations and found one rotting wooden bed frame under a cliff. A stove and miner's lantern were still there, along with many rusting tin cans and some broken glass. A good trail leads down the canyon and up to a slanting mine tunnel that was driven about seventy-five feet in the direction N 10° W, following the tilted layers of rock that contain thin veins of asbestos. Just south of the tunnel is a twenty-foot-high quarry face opened for a distance of 126 feet. Asbestos fibers up to four inches long were found in the mine. Much of the mining camp equipment was still present during one of Menkes' treks into the area in 1982.

We point out that since 1989 Hance's mine, prospects, and camp are off-limits to all visitation.

Figure 37. "Captain" John T. Hance in 1899. Photo courtesy of the late Otis (Dock) Marston, San Francisco, California.

As seen from the Colorado River at Hance Rapid, the rock layers of the Unkar Group rise up from river level in the downstream direction. Just below the rapids the Bass Limestone crops out, forming a natural ledge which a prospector could easily follow downstream, continuously rising higher from the river. The asbestos can be found where a black diabase sill has intruded the Bass Limestone. Hance's main mine was located a couple of miles downstream from Hance Rapid, high above the river. The nearest asbestos

From the campsite, prospectors followed the Tonto Platform westward into the wild tributary canyons on the north side of the Colorado. They located several claims in Clear Creek, Iron Canyon, and Dry Creek. The Iron Canyon and Dry Creek locations are not recognized today, but they may be tributaries to Vishnu Canyon.

The asbestos from the Hance Asbestos Company of New York (as he called it) contained fibers averaging two and one-half inches long, exhibiting a golden yellow color, shaded in places to a pale green. Hance's mine never made him much money. An unidentified person bought his claim for ten thousand dollars, which Hance spent in ten days in

San Francisco (Corle, 1946). He did somewhat better in the tourist trade.

In 1884, Hance had acquired squatter's rights to a ranch near his trail. There he operated a meager tourist business from his cabin and tents, professing himself (correctly, for a change) to be the premier Grand Canyon guide. He conducted tourist parties down his trail to Hance Rapid, and in this burgeoning service industry his name became synonymous with Grand Canyon. In 1893, a group of geologists from around the world, attending the 5th International Geological Congress (IGC) in Washington, D.C., traveled by train to Flagstaff where they transferred to stages for the long, jarring trip to Hance's hostelry on the Canyon rim. They were surprised by a storm and were forced to bivouac, and reached the Canyon the next day. Hance's guest register, now held in the Research Collections at Grand Canyon National Park, was partially reprinted by Woods (1899), and it provides the only known list of the IGC members who went into the Canyon with Hance—twenty-seven men and one woman, from England, Scotland, Germany, Austria, Switzerland, Belgium, France, and the United States. They enthusiastically and spontaneously renamed Hance's side canyon Congress Canyon, but the name never caught on.

In 1895, J.W. Thurber and H.H. Tolfree bought Hance's hotel and trail, and Hance returned to his asbestos mining of sixteen claims that were patented in 1901 (Verkamp, 1940, 1993:19, citing Coconino County Record of Deeds, No. 32, p. 599, County Recorder's Office, Flagstaff). These claims were subsequently bonded to the Hance Asbestos Mining Company, a Massachusetts company headed by George E. Hills, which company bought the mines a few months later for about $6,500 and placed Hance in charge of them. By 1904, Hance was removed as the superintendent, and eventually the mines were sold for taxes (Verkamp, 1940, 1993:19, citing a personal interview with R.P. Gilliland).

When Hance gave up his mining activities he managed a clapboard hotel owned by Lyman Tolfree. From there Hance would conduct parties into the Canyon, camping alongside the Colorado opposite the magnificent diabase dike in the Hakatai Shale. Near the mouth of Red Canyon, Billingsley in the mid-1960s found parts of tent poles, rock foundations, assorted pots and pans, tin cans, bottles, and an iron cook stove, indicating a sizeable campsite. These were probably the remains of Hance's riverside tourist camp, and by 1974 they had disappeared.

The remote clapboard hotel did not fare well against the comfortably rustic accommodations provided by the Fred Harvey Company at the railhead in Grand Canyon Village, and it was just a matter of time before the enterprise failed. But Hance was so well known by this time—people came to the Canyon just to see him—that he was taken on by the Fred Harvey people to do nothing but mingle with the guests, to tell tales, and to provide greater color than the Canyon itself. This he did until late in 1918, when he became ill. The Harvey Company kept him at the Canyon as long as they dared, then took him to the Weatherford Hotel in Flagstaff. After spending about a month at the Weatherford, Hance was hospitalized, and on the afternoon of January 6, 1919, he died, just weeks before Grand Canyon became a national park. John Hance rests today with other Grand Canyon pioneers in the cemetery at Grand Canyon Village.

Harnessing Water

Ever since Robert Brewster Stanton envisioned an electrically driven railroad through Grand Canyon, charged by turbines along the mighty Colorado River, various mining operations also thought to employ the Canyon's waters to electrify the mines. Even though the mine operators never went so far as to put the plan into operation, a commercial electric company did attempt to harness Bright Angel Creek. Julius Aubineau, David Babbitt, and A. Barman incorporated the Grand Canyon Electric Power Company on August 23, 1902. For two years they worked on the project, but, repeating the fate of many of the mines, the escalating costs of working in so remote a locale finally shut them down (Verkamp, 1940, 1993:20).

Nankoweap Canyon Area

Prospectors were active in the remote northeastern corner of the Canyon, in the Nankoweap Canyon area. The Nankoweap Trail was developed from an old Indian trail and made passable on horses in 1882 by a geological exploring party under the direction of John Wesley Powell, then the director of the U.S. Geological Survey. Charles D. Walcott, who became an eminent geologist and who later succeeded Powell as USGS director, spent the winter studying the strata and geological structure of the area around Nankoweap Canyon (Walcott, 1884). The Nankoweap Trail became a convenient route for prospectors.

Charles Gresswell, Bob Baumgarth, and others located the Pierce placer claims in 1891 on the west bank of the Colorado River, south of Nankoweap Canyon. It was the same story as the Kanab Canyon gold rush: flour gold too difficult to separate from the sand. Robert Baumgarth, with

J.D. Reese and Murray Averett, located several claims in October 1901 west of the Colorado River, on the north side of Nankoweap Butte.

In Nankoweap Canyon, a small mine called the Copper Mountain claim was started. All other claims in the vicinity were known as the Baumgarth claims. The late Proterozoic rocks here yielded very little ore and soon played out. A fifteen-foot-deep shaft at this location was inspected by Billingsley in 1969. Some picks, shovels, and drills still lay in the shaft, and were still present when inspected by Bob Dye in 1990. In the area are traces of the copper minerals malachite and azurite.

Toward Nankoweap Butte, marble-sized concretions of solid marcasite have been found weathering out of the Kwagunt Formation. This mineral is found in some of the ancient Prehistoric Pueblo sites along Nankoweap Creek (Robert C. Euler, written communication, 1990). Menkes found many tools in the Nankoweap area in the 1970s, but by 1982 they were gone. (One wonders at the motivation of hikers who will carry thirty-pound crowbars out of Grand Canyon.) Menkes searched for other claims between the mouths of Nankoweap and Kwagunt Creeks, but no traces were found.

Francis Mining District

Prospectors probably viewed Grand Canyon as a treasure trove once the word of silver and copper was out; however, not all of them took much solace in the potential payoff when the great efforts of getting in and out of the Canyon were taken into account. Some men stayed up on the Coconino Plateau to find their riches. Eventually, one copper deposit was discovered there, not far off the route between Williams and the Canyon.

Anita Mine

Around the time that the Orphan Lode Mine was discovered on the rim of Grand Canyon, another copper prospect was being developed fifteen miles to the south by William Owen "Buckey" O'Neill, one of the most colorful of the Canyon's entrepreneurial pioneers (for a biography see Walker, 1975). His mine and business foresight, probably more than anyone else's, indirectly spurred the growth of tourism at Grand Canyon Village.

On January 25, 1897, O'Neill began the first work on claims in the Francis Mining District. The Northern Star and thirty-three other O'Neill claims were clustered around a site called Camp Anita or, in some accounts, Copperopolis.

The exact date of discovery of the Anita Mine, which was initially called the Emerald Mine, is unrecorded, but in 1890 the Copper Queen twelve miles south of O'Neill's claims and other properties were already in the possession of a Mr. Nellis, a Williams, Arizona, pioneer. In July 1897 Nellis and Gerald M. Fennel also located several copper claims around Anita and near Red Butte to the east. Early on, it was believed that the copper deposits between Anita and the Orphan mine were part of an ore belt, which in turn was part of a greater belt that extended from Montana to Arizona (Williams News, July 17, 1897).

O'Neill persuaded the investment firm of Lombard, Goode and Company of Chicago, New York and London to put up development money, and they formed the Tusayan Development Company which bought up claims between Williams and Rowe Well. Rowe Well supplied water to Camp Anita, although a catchment had been built (Williams News, July 17, 1897). The most promising were Camp Anita, the Marshall D. McClure camp eight miles south of the Canyon, a camp at Rowe Well (where superintendent Timmons lived), and the Hogan and Ward camp at the Canyon. The most successful development work at Anita was at the Short-Cut claim.

Ore was hauled to Williams in a series of trailers attached to a steam-powered tractor which was fired with wood cached along the way. It seemed obvious that a railroad was needed for ore shipment and O'Neill and Lombard, Goode and Company, after being refused backing by the Santa Fe Pacific, raised $200,000 from local businessmen to build the Santa Fe and Grand Canyon Railroad. Surveys were completed and track laid to the copper mines at Anita by March 1900. Lack of funds prevented continuing track to the Canyon.

The Anita spur was "the little baby railroad born with a copper spoon in its mouth" (Williams News, October 14, 1899). Full ore cars did not even need a locomotive; it was downhill all the way from the mine to the junction. In actuality, the railroad spur was used very little to haul ore (Waesche, 1933).

O'Neill was elected mayor of Prescott, Arizona, in 1898 and sold his interests in the Anita mines to Lombard, Goode and Company. He joined Teddy Roosevelt's Rough Riders and was killed in Cuba on Kettle Hill on July 1, 1898.

The Anita mines were not successful enough to support a railroad and Lombard, Goode and Company and the local investors suffered great losses. The Santa Fe Pacific bought the railroad and completed track to the Canyon rim by September 1901. Its subsidiary, the Grand Canyon

Railway, carried passengers to the rim until 1968.

Ownership of the Anita mines property passed to William Nesmith of New York in 1912. On January 1, 1914, William H. and Grace Lockridge of Anita filed a claim on the "Nellis Mine" (the Anita Mine). Connected claims were owned by the Cardenas, Nyack, and Five Friends Companies (James, 1901:67). Other known claims in this district were the Blue Bonney Mine, Copper No. 1 Mine, Grand Canyon property, Packrat claim, and Tellstar claim (Bliss, 1993).

The Anita Mine itself is about three and one-half miles northeast from the Anita rail junction, on the south side of a small canyon through which the old highway to Grand Canyon passed. A log cabin is located nearby the mine. The mine is small, consisting of just one level, mainly a dugout tunnel driven south about eight hundred feet. Several short crosscuts have been made at intervals along the main drift. About four hundred feet south of the dugout entrance, buckets of ore were brought to the surface by a hand-operated hoist from a thirty-five-foot-deep shaft. The copper ore is concentrated in a stratiform breccia zone (not related to breccia pipes) and extends an undetermined lateral distance in the Harrisburg Member of the Kaibab Formation. Lockridge had indicated that the mineralized zone extended to a depth of 160 feet. Chalcopyrite is common, but most of the ores are carbonates of copper such as malachite and azurite. Gibson (1952) mentioned that copper carbonates and limonite disseminated in limestone and sandstone are concentrated in cracks and fissures with high silica contents. Some of the limonite is pseudomorphous after pyrite. Geologists and mine experts have found the copper deposits in limestone beds with no continuous ledges or deposits of appreciable size (Walker, 1975).

When George Wharton James visited the Anita Mine in May 1900, he saw "hundreds of tons of high grade ore" and "not less than fifteen hundred tons of average ore." He also noted that in February 1899 "a shipment of non-select ore" yielded 13 percent copper;

a later one-hundred-ton shipment contained 21 percent copper (James, 1901:68).

William H. Lockridge had a contract for thirty tons of ore per day. The mine produced this much for ninety days, but at no time could the mine have been called truly profitable. The work was done by hand by Lockridge and his family, employing two or three people altogether. Ore was shipped by rail to smelters in El Paso, Texas, and Hayden, Arizona. At other times, it was hauled by truck nearly two hundred miles to Kingman. A fifty-ton smelter was built at Williams to handle the ore, but it was never used. The last shipment was made in 1930 to Hayden (Waesche, 1933).

Rowe Well Claim

Sanford A. Rowe operated a livery stable and farm in Williams. In 1890, a Havasupai called Big Jim guided Rowe to the Canyon, where Rowe established a campground a few miles south of present-day Grand Canyon Village. Because water is scarce at the South Rim, Rowe dug a well at the camp and claimed the land by right of discovery (figure 38). Since he already had property in Williams, Rowe had to locate the well as a mining claim

Figure 38. Rowe Well and camp as it appeared around 1900, in a photo taken by Arion, or John R. Putnam. Photo courtesy of Seaver Center for Western History Research, Natural History Museum of Los Angeles County, California, photo no. 8597.

and mill site. He located the Lucky Strike, Little Mamie, and Highland Mary claims. But the only mineral taken from these claims was water. They were acquired by the National Park Service in 1956 (Hughes, 1967).

Warm Springs Mining District

Some prospectors who reached the rim of Grand Canyon probably wondered if the surrounding plateaus held more easily reached mineral deposits. Cattlemen, too, likely looked around when they were not running cattle to ranges on the plateaus. Mines were located, some of which have already been mentioned (the breccia pipes of the Hualapai Indian Reservation, and the stratiform deposits of the Anita Mine). The Warm Springs Mining District takes in part of the remote North Rim.

Apex Copper Properties

On the Kaibab Plateau several copper deposits were found at scattered localities southwest and northwest of the present village of Jacob Lake. Tainter (1947) referred to this cluster of prospects as the Apex Copper properties. Practically no records exist of the copper deposits, but several claims were filed in Coconino County, Arizona, on September 28, 1891.

One small copper mine was located near the head of Warm Springs Canyon by Andrew F. Freyerson in August 1885. A few prospectors of the high country during that time were Joseph Meeks, Brigham L. Young, J.A. Huntington, and George Adams. All of their claims were filed in the Warm Springs Mining District. Additional copper claims were filed on November 17, 1891, by J.S. Best, Harry McDonald , J. Dodson, R.M. Kennedy, and Elmer Kane. Numerous people whose names appear on the claims were partners or friends but probably never really got near the diggings.

Kane and McDonald were members of Robert B. Stanton's Colorado River railroad survey in 1889 and 1890. Later, they, along with William H. Edwards, took a boat trip sponsored by J.S. Best down the Colorado expressly to prospect (Lavender, 1985). The Salt Lake Herald on November 22, 1891, carried a lengthy story about the Best expedition, based on an interview with expedition members John Hislop and J.A. McCormick. They had remained quiet about their prospecting not so much for what they found, but because they had not found much.

It is likely that the copper deposits near Jacob Lake were discovered in the late 1870s, but the copper boom there did not begin until around 1890, and continued intermittently into the early 1900s. Hundreds of claims from the Buckskin Mountains (Kaibab Plateau) are listed in the Coconino County records in Flagstaff dating from 1891 to 1901 (Crampton, 1972).

In 1909, Frank Wynn and Nephi Jennings mined some fine specimens of uranium, from an open cut in their properties on the Kaibab Plateau. The radioactive properties of uranium were hardly known at the time, and the Washington County News of June 10, 1909, observed that "This metal is used in hardening steel and gun metal and is very scarce, only small deposits being found in Colorado." These uranium samples seem to be anomalous. Their source has not been located, and whether they indeed came from the Kaibab Plateau is in question. Such may be the case with many claims in the eastern part of Grand Canyon.

Many claims on the Kaibab Plateau were located in and around Warm Springs Canyon. Aquilla Nebeker and his partner V.M. Ryan worked mines there in 1900, about the time a small mining settlement called Coconino City was established about eight miles west of Jacob Lake where Warm Springs Canyon enters the lower Kanab Plateau country. Shortly thereafter the town name was changed to Ryan, by which the abandoned site is still known. A smelter was built around that time to refine the copper ore taken from the Jacob Lake and Warm Springs Mining Districts. (In this publication, we have merged the practically coincident two districts into one, called the Warm Springs Mining District.) The smelting operations continued until the price of copper fell in 1907 and work in the mines then ceased (Hall, 1975).

Dart Judd gave an account of his impression of Ryan during a visit with his family in 1905 (Cox, 1982:316): "When we went down that little hill north of Ryan, there at the mouth of Warm Springs Canyon, we could see the town all lit up like a Christmas Tree with what were probably the first electric lights any of us had ever seen. Ryan also had a narrow gauge railroad to carry the ore, a smelter, a flume and a four-inch wooden pipe to bring running water down from Big Spring." The railroad probably ran through Warm Springs Canyon.

The copper claims of the Warm Springs Canyon area are at an altitude of 7,900 feet. The Mackin and Petoskey groups of claims are separated by Warm Springs Canyon itself. Jacob Lake is immediately southeast of the Mackin group, and the Petoskey group is about three miles to the southwest, near Lambs Lake. The first major development work on these claims was done south of Warm Springs Canyon by the Petoskey Mining Company between 1900 and 1902. The company acquired five claims, built a steam-powered pumping plant in Warm Springs Canyon near a spring three miles from the mill site, and finished construction on a leaching plant. Before the plant became operational a fire destroyed it, bringing an end to the development (Tainter, 1947). In addition to the Mackin and Petoskey groups, other claims in the area were the Black Beauty, Kaibab group, Kennedy, Little Buck, South Phantom Nos. 1-6, Spotted Bull, Apex Copper, and Copper Queen, scattered over an area of about 1,700 hectares (Bliss, 1993).

In 1901, the Coconino Copper Company located a group of claims that were predecessors to the Mackin group north of Warm Springs Canyon. A small amount of ore was treated in a reverberator furnace at Ryan, but the smelter failed because of the high silica content of the ore. A plant was then built to use the Neill and Burfeind smelting process, probably the first attempt to leach copper by using sulfur dioxide as a leaching reagent. Around 1902, a patent was issued on the process, but the method failed because of the instability of the copper compounds. The plant later was leased to the Esmerelda Precipitation Company of Chicago, which also experimented unsuccessfully with the ore, and in 1907 the Buckskin Mountain Copper Company attempted experimental leaching processes. Eight additional claims were patented, and in 1916 a few tons of selected ore were shipped. A period of considerable stock promotion followed but little mining was done until 1928 (Tainter, 1947).

At that time the Saint Anthony Copper Company was organized to mine the ore with a steam shovel, and a railroad was built to Ryan's newly constructed one-hundred-ton blast furnace. Fire again destroyed the plant around 1929, ending operations after one thousand tons of 8 percent copper ore was shipped to the smelter at Garfield, Utah. The Los Angeles Exploration and Metal Company, Ltd., obtained title to the Saint Anthony holdings.

Still others attempted to mine the copper at Ryan, hauling it to a railroad in Marysvale, Utah. The meager production, estimated to generate just twenty-five thousand dollars, did not cover the cost of production and haulage.

The ore-containing units at the Apex properties are mostly in the matrix of white brecciated chert in bedding planes and fracture surfaces of limestone, impregnated with malachite and azurite; small amounts of earthy cuprite and chalcopyrite have also been seen. Silver and gold are present in very minute quantities. The ore has ranged from 2 percent to 40 percent copper, averaging 7 percent (Jenning, 1904). The deposits are found only on top of the Kaibab Plateau, where the rock layers are horizontal. Theoretical models have been developed by geologists in attempts to identify the mechanism of mineralization in this kind of deposit, but the models have not been successful (Bliss, 1993). One small prospect northeast of Ryan is located in dipping strata high on the south rim of Jacob Canyon where that canyon emerges from the Kaibab Plateau onto the Kanab Plateau. It is possible that this anomalous deposit is located on a breccia pipe.

These stratiform deposits are similar to those of Anita, south of Grand Canyon, and those of the Laguna Lake and National Canyon areas on the Hualapai Indian Reservation. All of these copper deposits found on the plateaus are contained in brecciated layers of the Harrisburg Member of the Kaibab Formation.

Little Colorado River Mining District

The eastern part of Grand Canyon is its grandest part— deeper, wider, and aesthetically more scenic than any other. This is so because here the great flex of the earth's crust, called the Kaibab Plateau on the north and the Coconino Plateau on the south, is cut open by the Colorado River and its myriad side canyons. It exposes a rich geological history and interesting geologic structures normally buried at great depth, that after a couple of billion years contributed to the hundred-year history of mining in Grand Canyon.

By coincidence, the mineral deposits that are mentioned here all occur in fault breccias where mineralized waters have percolated through the broken and shattered rock debris in fault planes. The principle of redeposition of minerals in these brecciated zones is basically the same as that in breccia pipes but, rather than cylindrical, vertical features, the fault planes are like highly inclined sheets whose widths are just a few feet. There are breccia pipes in this general vicinity, but many of them have poor surface expressions and are not well (if at all) mineralized, so they have not been prospected. There are pipes and collapse structures along the Little Colorado River valley east of Grand Canyon (e.g., Wenrich and Billingsley, 1986), but they are beyond the area covered by this publication.

McCormick Mine and Tanner Claims

Up and down the Colorado River corridor in the easternmost part of Grand Canyon, there is a series of prospects and mines, none of which are in breccia pipes. Despite the amount of activity, very little is known of the miners and their ore production or for that matter just when they were working their various prospects. Their histories are intimately commingled by loosely organized facts, undocumented hearsay and oral traditions, and dubious legends. The McCormick Mine is the most well known of these workings, and it is best to bring them all to attention under its heading.

Loosely dispersed in time and space as they were, the Little Colorado River prospectors had met on April 3, 1880, to form what they called the Little Colorado River Mining District. They drew the boundary beginning at the San Francisco Mountains near Flagstaff, then eastward to the Black Falls of the Little Colorado, northward to Willow Springs, westward to the "Cohonena Mountain" (probably Gray Mountain), and southward to the beginning point. It did not include the canyon of the Little Colorado River or its confluence with the Colorado River, nor any part of the eastern Grand Canyon area. Since no district has ever been made to incorporate the easternmost Grand Canyon and the adjacent Little Colorado River, we herein assign that area to the Little Colorado River Mining District. The original conveners in 1880 elected Seth Tanner as Chairman and W.H. French as District Recorder (Arizona Daily Miner, April 12, 1880).

Of all of the transient prospectors of this area, Seth Benjamin Tanner (figure 39) is at once prominent and obscure. Although there are various Tanner name places—Tanners Crossing, Tanner Wash, Tanner Trail, Tanners Well—the man himself is elusive. Seth Tanner was born in Warren County, New York, on March 6, 1828. An early Mormon pioneer, he came west in 1847 with his family, and settled near Salt Lake City, Utah. In 1850, Seth and his brother, Myron, went to seek their fortunes in the gold fields of California, where they apparently met with fair success. In 1852 they arrived in San Bernardino, California, and established a farm with two other brothers. In 1856, Seth left for San Diego to mine coal, exhibiting the restlessness and adventurousness for which he was known. In 1858, he returned to Utah, and in 1875 he and a party of other Mormon pioneers were called upon by Brigham Young to establish settlements along the Little Colorado River. Tanner settled at Moenkopi, Arizona, but later he moved to a lonely place along the Little Colorado now called Tanners Crossing, near present-day Cameron. All the while, Tanner's activities are vague, but we do know he was an outdoorsman who did not miss civilization. He was well known to the Navajos of the area, who, because of his burly physique and amazing strength, called him Hastiin Chush, "mature man," but which was anglicized or corrupted to "Mr. Bear" (Tanner, 1974). Seth Tanner is buried at Taylor, Arizona.

In 1889, Tanner and prospectors Fred Bunker and Lewis Bedlias improved an old trail that descended from the South Rim east of the present-day Watchtower at Desert View. The trail is today called the Tanner Trail, and its upper part now descends from Lipan Point (west of Desert View), joining the old trail at the level of the Redwall Limestone. The trail originally was known as the Bunker Trail (McKee's 1982 monograph on the Supai Group still referred to this trail as the Bunker Trail), and the Horse Thief Trail when rustlers in Utah drove their stolen animals from the North Rim, down Nankoweap Canyon, south along the Butte Fault to Lava Canyon/Chuar Creek, across the river, and up the Tanner Trail to the South Rim. (The shallowest place in the river during low flows in this area is at Mile 66. Billingsley speculates that this may be the rustlers' crossing.) Although no one was ever apprehended, the following incident was related by James (1901), who was told the story by miner and hosteler William Wallace Bass. It shows that the dangers of prospecting in Grand Canyon were not restricted to landslides, heat prostration, thirst, and starvation.

In March 1886, Bass with two companions set out to find the legendary lost gold mine of John D. Lee (see Lost Mines and Hidden Treasures). Leaving their horses at the Tanner trailhead, they descended into the Canyon and soon came upon a band of thieves driving eighteen head of horses with newly altered brands. Bass learned later that the horses had been stolen in Albuquerque, New Mexico. Not wishing to arouse suspicion, the "horse traders" and gold hunters tried to act friendly. After offering the traders some whiskey, Bass and party learned of important water sources on the other side of the river. One of the traders even said he was with John Wesley Powell's Grand Canyon explorers. Bass and his two friends reached and crossed the river but fearing for their horses, which they had left at the top of Tanner Trail, they gave up their search for Lee's mine. Not surprisingly, their horses had been taken.

Rustlers notwithstanding, along the Colorado River between Red Canyon and the Little Colorado River Seth Tanner located several claims in 1890. Billingsley, on one of several treks into the area, found a prospect pit in the Shinumo Quartzite in Seventyfive Mile Creek which may have been one of Tanner's claims. Other claims were found on copper veins cutting diabase sills higher up on the black cliffs (Cardenas Basalt) above Cardenas Creek and 75 Mile Canyon. These claims also may have belonged to Tanner, but this is speculation because no claim notices were found in the rock monuments there.

Tanner never did much with his claims, but one of them, at the mouth of Palisades Creek, did have a fair showing of copper minerals. On May 27, 1903, George McCormick took over Tanner's mine at Palisades Creek and called it the Copper Blossom Mine. Associates in the claim were Fred Hensing, George Bailey, A.T. Switzer, and E.O. McCoy. (McCormick had earlier extended the tunnel that had been partially worked by Tanner and Frank French, all of whom at one time thought the original short tunnel to be the lost silver and gold mine of John D. Lee. McCormick extended the tunnel four hundred feet into the canyon wall, and then in desperation drove a second tunnel in from the other side of a projection. He had finally abandoned the effort in 1895.)

It was either Tanner or McCormick who built a driftwood log cabin on the broad sand flat just below the Copper Blossom Mine; its remnants can still be seen. Butchart (1970:39) described the area: "The miners lived in a cabin made of driftwood, and the usual trash is still there. The mine shafts often have a few inches of mineralized water on the floor and the air is bad."

In 1970, Billingsley visited the McCormick Mine and found it partially caved in and encrusted with alkali salt and gypsum, with a few traces of copper minerals. On the flat, some of the cabin walls still stood. Menkes photographed the cabin in 1990. Spamer and Richard Quartaroli visited the mine in May 1994 and found the same encrustations. About twenty-five feet into the tunnel, cool, mineralized water flooded the floor. Outside, the canyon wall adjacent to the mine seeped mineralized water, leaving wet, orange and white oxidized encrustations on the rock surfaces. All the water percolates through the Palisades Fault.

The National Park Service has done some trail work in the mine area. Clear pathways direct visitors from stumbling into and destroying the remnant walls and trash piles.

The mineralized zone in the Palisades Fault betrays the mode of emplacement of the copper-rich ores of the McCormick and other mines in the fault zone on both sides of the Colorado River. Faults naturally allow ground water from rain and snow or highly mineralized waters from deeper in the earth to pass through rock layers. The time of mineralization in this fault is not determined, but it is likely that it was about the same time as when the breccia pipes of the southwestern Colorado Plateau were mineralized. Mineralized fluids migrating through the strata probably also interacted with the Proterozoic basement, in which the Palisades Fault occurs. (In another example, yellow carnotite was redeposited in the Bright Angel fault zone in Bright

Figure 39. Seth Benjamin Tanner (1828-1918), Grand Canyon pioneer and prospector, in a photo taken between 1900 and 1910. Photo courtesy of George Tanner.

Angel Canyon. See notes to Appendix 2.) Of course, at the time the area was deeply buried. The Colorado River and its Grand Canyon still were far in the future.

Mines in the Chuar and Basalt Canyon Areas

The Tanner Trail was the principal route down which prospectors reached the scattered claims in the eastern Grand Canyon. It also is the way to the only ford on the river which is easily accessible to the several routes off the rims in this area and is the same ford that is by the McCormick Mine and across which the horse rustlers drove their ill-gotten wares. Across the river from Palisades Creek, Grand Canyon photographers Emery and Ellsworth Kolb, on their motion picture filming of a trip down the Colorado River, took note of the mine workings in this area in December 1911. They observed "some tunnels on both sides of the river where the Mormon miners had tapped a copper ledge" (Kolb, 1914:204). Whether or not they stopped at the mines is unclear, since they were pressed for time that day.

Prospector George Bailey did not have to stray far from the Mile 66 ford. Immediately opposite "Loxena Creek" (Palisades Creek) and the McCormick Mine, Bailey made the Copper Grant claim on January 4, 1904, in the same brecciated zone of the Palisades Fault in which the McCormick Mine was located. Although he refiled the claim on January 4, 1914, the mine was not a good producer. There is a mine tunnel alongside the Colorado River at the mouth of Lava Canyon; location is shown in Hereford's (1993) pl. 1. It follows the fault fifty feet or so before making a right-angle turn to the right and driving in about another thirty feet, and is still partly shored with timbers. There also are more extensive diggings farther up the hill in the Palisades Fault. Any (or all?) of these prospects may be the Copper Grant.

Harry McDonald filed the Morning Star claim in "Silver Creek" (Chuar Creek in Lava Canyon). The deposit was discovered in May 1890. The mine entrance is completely caved in and very little copper ore is found at the site although a small piece of native copper was found in the summer of 1989 by participants of a geology expedition down the Colorado River.

Others involved in claims in this area were J.N. Hughes, E.A. Reynolds, Al Peterson, Nels Johnson, and Al Huntington.

In 1889 and 1890, an ambitious expedition headed by Robert Brewster Stanton was methodically floating down the Colorado River, surveying a route for a railroad between the Rocky Mountains and the Pacific Ocean (see Protection of the Grand Canyon). The party took note of mineral occurrences but they apparently made no claims after they passed Lees Ferry and entered Grand Canyon. This is curious, considering that they recorded glowing accounts of mining riches in the Canyon from a prospector they met on January 21, 1890, in the Furnace Flats reach of the Colorado where it meanders through the open spaces below Desert View.

They had met Felix Lantier (figure 40), a prospector from Flagstaff. He spent a couple of days with the party and told them about mineral resources he knew of in the area (Smith and Crampton, 1987:155-156). He spoke of gold, silver, and copper "near Powell's volcanic dike" (at Hance Rapid), and of other sources yielding lead carbonate ore and blue roofing slate. Stanton said he had been given

samples of all of the rocks except the roofing slate, but the fate of these specimens is not known. The party pressed Lantier into service as a postman, and he departed on January 24 with valuable survey notes, photographic negatives, and letters which he delivered safely to the Flagstaff post office (Stanton, 1965:146).

When the Kolbs passed the Tanner and Hance Trails in 1911, they had looked for signs of prospectors, which indicates that some straggling prospecting activity was continuing in this area (Kolb, 1914:208-209).

A half-hour's run and a dash through one violent rapid [Hance Rapid] landed us at the end of the Hance Trail—unused for tourist travel for several years—with a few torn and tattered tents back in the side canyon down which the trail wound its way. We half hoped that we would find some prospectors who make this section their winter home either at the Tanner or Hance Trail, but there was no sign of recent visitors at either place, unless it was the numerous burro tracks in the sand. These tracks were doubtless made by some of the many wild burros that roam all over the lower plateaus in the upper end of the Grand Canyon.

Figure 40. Felix Lantier and his dog, in January, 1890, near Colorado River Mile 69 (Tanner Rapids area). Photo courtesy of the late Otis (Dock) Marston, San Francisco, California; from the U. S. National Archives.

In 1928, Grand Canyon National Park's naturalist Glen Sturdevant, chief ranger James P. Brooks, ranger Arthur L. Brown, and packer Jack Way descended the Tanner Trail to reconnoiter the northeastern section of the park. After following the Colorado upstream, heading for the Mile 66 ford, they came across some old mining sites opposite the mouth of Chuar Creek. Sturdevant (1928) reported seeing "Standing frames of tent houses and an old boat showing that active mining operations were once carried on. The sides of one prospect tunnel [the McCormick Mine] encrusted with carbonate of copper and salt." They found another camp, which they first thought to be an abandoned mining camp, in a tributary to Chuar Creek. It proved to be "a complete distillery which was probably used by some prospector to continue business with his poor-paying mining operations." The operators of the still are unknown. A wrangler for a 1933 U.S. Geological Survey party in Nankoweap and Chuar Canyons told one of the geologists that rustlers had built the still (William Swayne, personal communication to Menkes).

John Hendricks (personal communication, 1988), of the U.S. Geological Survey, mentioned that in 1971 he had seen several old copper mines in Basalt Canyon (about two miles south of Chuar Creek), along with some mining tools. Near these mines, in what the miners called Marshall Canyon, Donald P. Elston of the U.S. Geological Survey found a mining claim for the Marshall Lazure Group Mine, "situated and located in the unnamed Mining District," dated January 1, 1915, recorded by Charles E. Lazure and witnessed by J.W.B. Calhoun and Elliott Hodgen. That claim is located on a fault in Basalt Canyon.

Prospects on the Little Colorado River

Ben Beamer lived at the confluence of the Little Colorado and Colorado Rivers, where for a house he had rebuilt a Prehistoric Pueblo ruin under a ledge of the Tapeats Sandstone, a short distance up the Little Colorado on the south side (figure 41). Beamer had intended to homestead the banks of the Little Colorado, for which he brought in such articles as a stove, cooking utensils, and frying pans. A plow found there very likely was his. All of these items were still inside the cabin in 1967. Some items have disappeared since then, but many relics were seen by Spamer in 1990 and 1991. Most astonishing is the idea that Beamer would think of farming here. The banks of the Little Colorado are sandy, the water is full of magnesium salts (or runs as a muddy, chocolate-colored river during spring and storm-produced floods), and a more lonely place cannot be easily found.

The Denver Republican of July 17, 1892, ran an account by Ben Beamer, which is probably our best portrait of the recluse:

I got into the canon by the Tanner trail in February 1890. I have lived there ever since, except for short trip to the outer world last winter, and during the whole time I have been there, I never saw a human being until this spring, when line surveyors of the Atlantic and Pacific Railroad, with a guide, made their way into the canon. I took up a ranch at the mouth of the Little Colorado, where there are about ten acres of cultivatable land, built me a cabin and went about my own business of prospecting for the precious metals. All about there are strong indications of copper, it being a sandstone country underlaid with shale, a No. 1 copper formation. Six miles below me, on the south side of the Colorado, is the McDonald claim, which belongs to Denver parties, J.N. Hughes, the lawyer, being one of them, I believe. There is from 60 to 70 percent copper in the rock, and some silver. Below that Mr. Hance and two partners have some big copper claims, and still farther down a man named Berry has located some claims which show the same percentage of copper and run from $10 to $100 per ton in silver. An Asbestos claim below Hance's was lately sold for $7,500. How do I live? Well, as all prospectors do, only I get plenty of fish and wild goat, and there are some otter. After the snow melts the Colorado backs up into some of those small canons and the fish come in millions to feed on a vegetable that grows on the rocks. They are so thick that you can lean over the water's edge and pull them out by the tail two at a time. Facts, I assure you. No, it's only in the Little Colorado where they cannot live. They are about twenty inches long and have a flat hump on their back just behind the head. Lonesome? Not when you get used to it.

Beamer established a trail between the Tanner Trail area and the confluence, probably following the same trail used by ancient peoples to reach the salt deposits. Today, this trail is called the Beamer Trail. (The flat-humped fish, incidentally, are the humpback chub, *Gila cypha*. The fish and its habitat are federally listed as endangered. Karen Underhill (personal communication to Menkes) has pointed out that the newspaper article contains the first record of the chub in the Canyon not noted by biologists in their records. Previously the earliest records have been attributed to the National Geographic article by Kolb and Kolb, 1914, and Ellsworth Kolb's book about their journey, 1914.)

Figure 41. Ben Beamer's cabin as it appeared in July, 1967, on the south bank of the Little Colorado River, under a ledge of Tapeats Sandstone, a short distance upstream from the confluence with the Colorado River. Photo courtesy of Vernon A. Taylor, Prescott, Arizona.

Very little is known about other, scattered claims in the Little Colorado River area. Charles Spencer and a fellow miner, Dan O'Leary, staked a claim (probably a placer claim) near the mouth of the Little Colorado in December 1877, but nothing ever came of it (Weekly Arizona Miner, December 14, 1877). During 1891, Ed Gale, R.H. Cameron, and W.G. Steward made several claims, presumably for placer gold, one and one-half miles east of the Little Colorado and near the mouth of this stream. Vernon Taylor, a geochemist from the University of California at Los Angeles, while researching travertine deposits in the confluence area in 1968, found a mining claim dated July 4, 1929, near the junctions of Big Canyon and Salt Trail Canyon within the Little Colorado River. The name was not legible, but it was a placer claim for gold.

Akens' (1991:16-17) map of uranium prospects in the Four Corners, Grand Canyon, and eastern Nevada regions, originally drawn by Ed Smart, shows a string of ten prospects along the Little Colorado River, from the gorge west of Cameron southeastward to Holbrook. The text, however, does not discuss them.

Lees Ferry Area

The Lees Ferry area of Arizona is not in any designated mining district. As an important Colorado River crossing,

near the mouth of a major regional tributary, the Paria River (Rusho and Crampton, 1975, 1992), it saw its share of transient prospectors, and they sought gold even here. There was some mining activity in the area by early 1879 when the a Salt Lake City newspaper reported only that "A mining camp has been established on the Pahreah, forty miles east of this place [Kanab]. The quality of the ore is not publicly known. Great excitement exists among the miners, but no reliable information has been comeatable [sic] until the present" (Salt Lake Tribune, January 21, 1879). The mining camp sprung up around "Ketchum's gin mill," later called Micawberville, the seat for two hundred disappointed prospectors in the area. The district was called Silver Bow (Salt Lake Tribune, February 7, 1879). As one anonymous "Interviewer" reported, ". . . suckers run early this spring, especially on the Pahria" (Salt Lake Tribune, February 9, 1879).

In 1897, Julius F. Stone, with Robert B. Stanton, formed the Hoskininni Mining Company to exploit placer gold deposits along the Colorado River in Glen Canyon, Utah (now inundated by Lake Powell). Dam and power sites were surveyed, and 145 placer claims were located (Spence, 1993). They assembled a huge, 180-ton dredge containing five gasoline-powered engines and forty-six three-cubic-yard buckets, which succeeded in dredging gold so fine that it washed right through the machinery. They recovered $66.95 worth of gold and went bankrupt in 1902 (Webb, 1994).

Charles H. Spencer arrived at Lees Ferry in May 1910. Gold was known to occur in small quantities in the green shales of the Petrified Forest Member of the Chinle Formation, but Spencer saw more gold in his dreams than actually existed. He set up a system of hydraulic hoses to shoot Colorado River water at the shale slopes a few hundred yards away. The dislodged and dissolved material flowed back down a long flume toward the Colorado River and an amalgamator (figure 42). In the spring of 1911, the first runs of shale were made.

Meanwhile, Spencer and his crew built a trail (now the Spencer Trail) up the cliff behind the operation in order to bring coal to the placer operation (The coal was blasted out of exposures in Warm Creek Canyon; Kolb, 1914:172). After some trial and error, a boat, the Violet Louise, was used to tow a coal barge, but this practice was short lived. The boat did not have the power to tow even the empty barge upstream.

To remedy the fuel situation, a steamboat was ordered built in San Francisco. It was shipped in pieces by rail to Marysvale, Utah, then hauled by wagon to the Colorado River twenty-eight miles upstream from Lees Ferry. Near Cedarville, Utah, "the wagon carrying the boiler missed a turn. The boiler rolled seventy-five feet downhill, knocking over a grove of cottonwood trees. It took thirty-two days to get the ten foot long, one hundred horsepower marine boiler back on a wagon" (Avery, 1981:7).

Figure 42. Sluices at the Charles H. Spencer gold placer operation at Lees Ferry, around 1910, leading back toward the Colorado River, and the amalgamator shed. Photo reproduced with permission of the Huntington Library, San Marino, California.

On November 3, 1911, Emery and Ellsworth Kolb, on their historic filming run of the Colorado River, rounded a turn in the river only to witness

> *…the strange spectacle of fifteen or twenty men at work on the half-constructed hull of a flat-bottomed steamboat, over sixty feet in length. This boat was on the bank quite a distance above the water, with the perpendicular walls of a crooked side canyon [Warm Creek Canyon] rising above it. It was a strange sight, here in this out-of-the-way corner of the world. Some men with heavy sledges were under the boat, driving large spikes into the planking… (Kolb, 1914:171).*

Construction of the Charles H. Spencer was completed on February 12, 1912. No one was particularly experienced with handling such a large vessel, and "A cowhand named Pete Hanna captained the boat by dint of some experience on the Mississippi" (Avery, 1981:9). After loading the hold and decks with coal, the novice navigators steamed on down the Colorado, with a mishap or two, arriving at the ferry stern-first. "Captain" Hanna, however, refused to unload the coal; it would take all of it for the return trip twenty-eight miles upstream. So the Violet Louise's abandoned barge was pressed into service, towed upstream (the Charles H. Spencer indeed consuming all the coal), reloaded with fifteen to twenty tons of coal, released and piloted freely with barge poles back downriver to the ferry, the Charles H. Spencer following behind, stern-first (Avery, 1981).

Everything worked as well as could be expected for a while, until the mercury in the amalgamator became clogged; then the gold dust washed away with the tailings. Numerous unsuccessful efforts were made to solve the problem. Samples were sent to Marie Curie in Paris, who indicated that the amalgamation process had failed because of the presence of an indeterminate element (Avery, 1981). Troubles continued as the company failed to meet its payroll, and liens were placed on the properties (Kane County Independent, April 11, 1912). The placer operation ended in 1913.

Spencer eventually learned that the element that caused the amalgamator to clog was a rare metallic element, not identified until 1925, called rhenium. When a method for removing rhenium was developed, in the 1960s Spencer, then ninety years old, again started the hydraulic mining technique at Paria, a small settlement upstream along the Paria River from Lees Ferry. But this time he was after the rhenium itself, a highly valuable superconducting metal useful in specialized electronics. This venture failed, too, and Spencer said his success was blocked by international mining cartels that did not want his vast ore deposit on the market. He never did strike it rich (Rusho and Crampton, 1975).

The original Spencer gold operation cost about $350,000, offset in part for one co-investor, William Switzer, when Spencer accidentally left a half-full jar of gold dust in his Flagstaff office (Avery, 1981). The Charles H. Spencer was abandoned at Lees Ferry and sank in 1915. Today, a few rusted remnants, including part of the rusted hapless marine boiler, remain on the west bank of the Colorado.

CONTINUING MINING ACTIVITY

The historical component of this publication has addressed several distinct mining districts. Modern mining activities and mining law have less heed of specific districts than did early claim records, so we here separately discuss the mining activities that are continuing to this day in the Grand Canyon region.

Oil Exploration

The prospect of oil beneath Grand Canyon's plateaus has not been overlooked. The first oil well drilled near the Canyon was on the Arizona Strip in 1906. As of 1961, eighteen wells had been drilled but none were productive; they were classified as dry wells and abandoned. Six wells north of Tuckup Canyon showed numerous oil stains in the Toroweap Formation and in the base of the Coconino Sandstone. Many of the wells bottomed out in the Redwall. Four deep exploration test wells, ranging from 3,753 to 4,666 feet in depth, have been drilled in northern Mohave County. Light oil stains were found in the lower Toroweap Formation, Coconino Sandstone, Callville Limestone, Redwall Limestone, and Muav Limestone. These wells, all commercially marginal, were plugged and abandoned (National Park Service, 1977). Giardina (1979) prepared a review of geologic features and structures of the Grand Canyon region, including a list of nineteen northern Arizona wells drilled for oil and natural gas, for the use of petroleum investigators (Table 3).

No Arizona Strip oil well was less than twenty-five miles from the Colorado River. A few holes, which yielded nothing, were drilled south of Grand Canyon. The deep erosion by the Colorado River and tributary canyons probably exposed and allowed whatever little oils ever existed in them to leach out from the rocks.

The Coconino County mining records (Book 9, pp. 34-37) list several oil claims in the Lees Ferry area, requested by E.H. Spencer on January 27, 1914. Listed on the claims were Herbert A. Parken, J.H. McFarland, Rolla E. Clapp, H.M. Huttey, A. Walles, W.H. Switzer, and A.M. Turner.

S.S. Acker and C.E. Code located several claims around the Little Colorado River area in April 1916. None of these claims produced any oil of quality.

Recently, the potential for oil has brought attention to the eastern Grand Canyon area. Reports by Summons et al. (1988) and Reynolds (1989) indicated that some rock strata of the Proterozoic Chuar Group contains as much as 5 percent total carbon, putting these rocks within the principal oil-generating window. These authors suggested that the rocks still have potential to produce sufficient amounts of gaseous and liquid hydrocarbons for commercial use. The strata, except where they are exposed in Grand Canyon, lie at depth beneath the Kaibab Plateau and parts of southern Utah. Some oil companies have planned to drill exploratory wells in areas of the Marble, Kaibab, and Paria Plateaus north of Grand Canyon. Around 1991, a wildcat well was drilled about six miles west of Fredonia to look for Proterozoic oil.

Uranium Booms

There were two uranium booms in the Grand Canyon region. The first was in the 1950s, inspired by the discovery of uranium oxide ores across the Colorado Plateau, locally in the Hacks Canyon and Orphan Lode mines, as well as the escalating price of uranium. The second boom was in the late 1970s when skyrocketing uranium prices spurred on the exploration for mineralized breccia pipes on the Colorado Plateau.

The uranium fever of the 1950s (see, for example, Ringholz, 1989) spread to the Four Corners area and the valley of the Little Colorado River. There, the Shinarump Member of the Chinle Formation, a basal conglomeratic sandstone, contains pockets of organic materials such as carbonaceous fossil plant debris and petrified wood. The Shinarump and the overlying Petrified Forest Member were found to contain rich pockets of uranium ore associated with the organic material in the area around Cameron, Arizona. The geological nature and ore potential of breccia pipes were unknown at the time, awaiting the detailed studies made during the development of the Orphan Lode Mine in the late 1950s (Chenoweth, 1986).

Charlie Huskon, a Navajo prospector employed by the Atomic Energy Commission, made the first uranium discovery in the Chinle Formation near Cameron in 1952 (Chenoweth and Akers, 1958). The discovery led to similar finds near Lees Ferry, Holbrook, and St. Johns, Arizona. The uranium ore occurs in very localized pockets, and as a result many small open-pit mines soon pockmarked the Painted Desert around Cameron. The operations were short lived, but the mining scars will be visible for decades to come.

Blue Mountain Claim

The rich uranium ores found in breccia pipes created a frenzy of exploration on the southern part of the Colorado Plateau during the late 1970s and early 1980s. By this time, their potential had been well established by the Orphan Lode, Hacks Canyon, Chapel House, Ridenour, and Blue Mountain Mines.

The Blue Mountain pipe had been known for years, but the first recorded examination of this deposit was in 1953 by R.D. Miller as part of a uranium reconnaissance program conducted by the Atomic Energy Commission (Miller, 1954a). The Blue Mountain pipe is located at the western edge of the Coconino Plateau, on the southeastern part of the Hualapai Indian Reservation, a few miles east of Diamond Creek. The claim is in a small tributary drainage to Diamond Creek and forms a small but prominent spire jutting up from the Coconino Sandstone. It is composed of silica-cemented, brecciated and fractured Coconino Sandstone, with copper minerals, mainly malachite, azurite, and chrysocolla, occurring as splotches in the cement (Van Gosen et al., 1989). Western Nuclear leased the pipe from the Hualapai Tribe from 1976 to 1978 and conducted an exploratory drilling program. Uranium ore was found only in thin, intermittent horizons, the most prominent one about three hundred feet below the surface. Not enough enthusiasm was generated to begin a mining program, and the lease expired without production in November 1978 (Van Gosen et al., 1989). In 1984, Karen Wenrich of the U.S. Geological Survey headed a project to drill the Blue Mountain pipe again at the request of the Hualapai Tribe, to scientifically study it and to redetermine the uranium potential. Although the program was scientifically successful, the economic potential was as disappointing as it had been declared to be by Western Nuclear.

Western Nuclear also did some exploratory drillings in September 1976 into Tertiary-age gravel deposits in an eastern tributary to Peach Springs Canyon. These are known locally as the Mulberry Spring uranium prospects (Karen Wenrich, personal communication, 1987). Gamma radiation at this locality is nearly six times that of background values measured in nearby outcrops of Cambrian limestones and shales; however, no mining has been done here.

Besides Western Nuclear (later Energy Fuels Nuclear), claims were made across much of the plateau region adjacent to Grand Canyon by Pathfinder Mines Corporation of St. George, Utah; Uranez USA of Denver, Colorado; and Rocky Mountain Energy (later Union Pacific Resources—Minerals) of Denver. Pathfinder Mines located two uranium ore bodies in breccia pipes west of Hacks Canyon, the EZ-1 and EZ-2. Union Pacific Resources found two breccia pipes that contain uranium ore: the Sage pipe, about ten miles south of Supai, on the Coconino Plateau, and the SBF pipe farther to the southwest, near the eastern boundary of the Hualapai Indian Reservation. As of 1990, no mining had begun on these claims.

Canyon Mine

The Canyon pipe, about twelve miles southeast of Grand Canyon Village near Red Butte, was discovered by Gulf Mineral Resources in late 1976 using aerial photographic methods. The ore itself was confirmed by exploratory drilling in May 1981. The Canyon pipe was sold to Energy Fuels Nuclear in the mid-1980s, and development of the mine began in 1987. However, during the 1980s environmental groups began to protest some of the mining activities near the Grand Canyon. They considered the uranium mining a potential contributor to the pollution of ground water used by the Havasupai Tribe in Havasu Canyon and elsewhere. Permits and official clearances, in this case from the forest service, became increasingly difficult to obtain, and a long list of grievances from various environmental groups delayed mining here and at other sites in the region.

Meanwhile, uranium prices declined steadily from the 1979 peak. This, together with legal and political uncertainties about the future of nuclear power, foreign competition plaguing the domestic uranium industry, and eventually the stripping down of the nuclear arsenal after the unforeseen collapse of the Soviet Union, caused Energy Fuels Nuclear to cut back operations here and at other mines on the Kanab Plateau. The Canyon Mine, however, continues to be worked, and its expected life is five to seven years. Afterward, the company will remove its equipment and reclaim the land.

Other New Breccia Pipe Mines

Energy Fuels Nuclear, being the first to excel at detection techniques for locating breccia pipes on the plateau surfaces, made several claims for uranium north and south of Grand Canyon. The Pigeon pipe, near the north rim of Snake Gulch, a tributary to Kanab Canyon on the Arizona Strip, was discovered in early 1980. Exploratory drilling on this pipe in April 1981 revealed that a significant uranium ore deposit exists at depth. In December 1984, the Pigeon Mine was begun. The ore was finally depleted by 1991 and the mine closed.

Not far to the west and southwest of the Pigeon Mine, four breccia pipes found by Energy Fuels Nuclear were discovered to contain high-grade uranium ore. The Kanab North Mine, located about the same time as the Pigeon Mine in late 1980, is on the west wall of Kanab Canyon. Exploratory drilling in September 1981 revealed uranium ore, and mining began in December 1987. The mine closed in 1991 when uranium prices dropped.

Farther west on the Kanab Plateau, the Hermit breccia pipe was drilled in October 1982, and by late 1988 the Hermit Mine was being worked.

South of the Hermit Mine on the Kanab Plateau, the Pinenut pipe was discovered and drilled in August 1985. Mining for uranium began there in March 1988.

The Arizona 1 pipe, also on the Kanab Plateau about five miles southwest of the Hacks Canyon Mine, was drilled in November 1986. Although uranium ore was found, no development began until 1988, when six uranium ore bodies were being mined. They produced 2.5 to 7 million pounds of ore, yielding .5 percent to .7 percent uranium oxide (personal communications from personnel of Energy Fuels Nuclear, Kanab, Utah, 1988).

Limestone

The most abundant minable resource of the whole Grand Canyon region is the rock that makes up most of the surface of the plateaus: limestone. The Nelson limestone quarry near Nelson, Arizona (figure 16), is operated by the Gemstar Lime Company. It is the longest-operating mine in the Grand Canyon region and continues to produce high-quality limestone from the Mooney Falls Member of the Redwall Limestone. It is processed to manufacture cement and road materials.

Potential Prospects

Stratiform copper deposits of the Warm Springs and Francis Mining Districts are well prospected insofar as surface exposures have allowed. It is not likely that new deposits will be found in the Warm Springs district, but the Francis district has adjacent to it large areas of lava flows that were derived from the San Francisco volcanic field. These rocks cover the Harrisburg Member of the Kaibab Formation, and perhaps some copper deposits as well. A large part of the area is administered by the Kaibab National Forest, but it is not clear at this time what limitations or incentives may exist to search for developable prospects there (Bliss, 1993).

The lava flows of the San Francisco volcanic field may cover breccia pipes. Undiscovered pipes are likely to be scattered across the region of northwestern Arizona that is underlain by the Redwall Limestone. Perhaps half of the undiscovered uraniferous deposits may lie in the North Kaibab Ranger District of the Kaibab National Forest (Bliss, 1993).

chapter five

LOST MINES
& HIDDEN TREASURES

The lore of the "Lost Mine" is an integral part of the mystique of the American West. And the Grand Canyon has its share of lost mines and tales of hidden treasure. The first to fall prey to the legends were the Spanish conquistadors. The treasure ships of plunder from Mexico, taken by Hernando Cortes in 1520, spurred other expeditions of economic and spiritual salvation. In 1528, the de Narvaez expedition landed in Florida and failed. One of the few survivors, Alvar Nunez Cabeza de Vaca, wandered westward for eight years. Reaching Mexico in 1536, he told about fabulous cities reported to him by the native peoples, perhaps one of the first American instances of natives telling treasure seekers what they wanted to hear. This eventually led to the expedition under Francisco Vásquez de Coronado, in 1540, which sought (and laid to rest) the legend of the Seven Cities of Cíbola; the golden cities were the Zuni towns of New Mexico.

Lost Mines of the Padres

The Viceroy of New Spain heard a rumor of a rich metal, perhaps quicksilver (mercury, useful in separating gold from ore) to be found west of the Hopi pueblos in a "red hill at the foot of blue mountains." In 1691, he instructed Diego de Vargas, the colonial governor of New Mexico, to investigate. Vargas's inquiries turned up people who had heard the reports and said the location was "near a large river on the side of a red mountain." Vargas traveled west to the Hopi Mesas and obtained and took to Mexico a heavy, greasy, red ore which contained no mercury. Colton (1940) believed that the source was the red ocher mine in western Grand Canyon, described in this monograph as red pigment deposits. The location of the mine has never been determined.

The Lost Waterfall Gold

"Long Tom" Watson, a lone prospector who wandered northern Arizona, one winter settled into a shack in Flagstaff (the year is unknown). Among the newspapers he used to start fires, he found bundles of old letters. One day

as he was about to put a batch of paper into the stove, he noticed a letter that had not been opened. The writing on the envelope was illegible, but inside was a piece of brown wrapping paper with a note on one side and a map on the other, apparently a message from a prospector to his brother.

The prospector had found gold in Grand Canyon. Aware that two men had been following him, he filled a sack with nuggets and hid it in a small cave behind a waterfall. The next morning, the two men came into his camp, a gunfight ensued and the prospector was seriously wounded. He managed to drag himself to the Canyon rim, where he was found and carried to a doctor in Williams. There, the dying man wrote the tale for his brother in a letter that was never mailed.

Watson set off in the spring to find the sack of gold. He searched the major tributary canyons and spent a few years looking in Havasu Canyon because of its waterfalls. His search carried him upriver and eventually he reached the Tanner Trail. By now, Watson had deduced that the waterfall he was looking for probably flowed only in the spring. On the Tanner he heard the sound of water and was surprised to see a waterfall flowing from the wall beside the trail. He climbed to a ledge and plunged through the water. Sure enough, there was a small cave and in it a deteriorated sack full of golden nuggets. Watson filled sacks and his pockets with as much gold as he could carry and lurched back through the waterfall, but he lost his footing, fell, and suffered a broken leg. He came to the next day, hauled himself onto the back of his burro, and rode to the top of the Tanner where he was found and carried to a hospital in Flagstaff.

Four months later Watson shared his secret with "Doc" Scanlon, a veterinarian, and together they returned to the Canyon but were unable to find the ephemeral waterfall. Watson, bitterly disappointed, eventually took his own life with his rifle near the Canyon rim (Richardson, 1953). There is no record of what happened to the gold nuggets.

John D. Lee's Lost Mine

The legend of John D. Lee's gold mine is the Canyon's

most enduring. Though gold was hard to come by along the Colorado River, Lee supposedly found all he could carry in a secret location.

Lee, who was born in Illinois in 1812, came to Salt Lake City with other Mormon pioneers in 1848. He was a rugged and resourceful outdoorsman, tempered by the rigors and travel of western settlement. He is most well known for his part in the planning and execution of the massacre of 140 non-Mormon pioneers in Mountain Meadows (Utah) on September 11, 1857. He evaded arrest for twenty years, partly because his church sent him to the remote Colorado River crossing site now named for him, but was tried, convicted, and executed in 1877.

Lee supposedly explored a large sector of northern Arizona during the 1860s while he was in hiding, and lived for a period with the Havasupai in Havasu Canyon. Legend has it that during his explorations he found a rich vein of silver and gold near the confluence of the Little Colorado and Colorado Rivers. In Salt Lake City, during 1872 and 1873, he sold silver and gold which he had packed out of the upper section of eastern Grand Canyon. Lee never let anyone know where it came from and made sure no one followed when he wandered off into the Canyon. A boy named Robert B. Hilderbrand was the only person known ever to have been along on these excursions. Robert was left in camp when Lee went to the mines, to return a few days later with sacks of silver and gold (Kildare, 1975). The St. George newspaper, The Union, reported on March 6, 1897, that from the vicinity of Fredonia, Arizona, "Indians sometimes bring out from the canyon's depths specimens of rich gold and copper ore, and John D. Lee . . . is known to have found, when he was in hiding in the town below Lee's ferry, a large quantity of coarse gold."

Stories of Lee's lost mine were prevalent in the late 1800s, and its location was described variously by writers and reporters as being on either side of the Colorado River, somewhere along the upper part of Glen Canyon in Utah, and as far downstream as Lava Falls (Colorado River Mile 180). George Wharton James did much to spread the tales (James, 1901). He had heard of the mine from William Wallace Bass who, in turn, had learned of it only by hearsay. It was not until after Lee's diaries were published in 1955 that his whereabouts during the period in question were known with some certainty (Cleland and Brooks, 1983).

After Lee's death prospectors began to look for his secret mine. Among the hopeful searchers were Seth B. Tanner and partner Frank French (see the McCormick Mine). They found an old tunnel along the river at Palisades

Canyon and concluded that it was Lee's mine, but abandoned their work because they found nothing of value.

John Hance, of course, said he knew where the mine was, but by sheer coincidence it had been covered by an earthquake-induced landslide. Hance said he had met Lee in the Canyon several times. However, it is generally accepted that Hance came to the Canyon after 1883, and Lee was executed in 1877.

In 1890, a Mr. Brown from Utah met William Wallace Bass in Williams and told of Lee's lost gold. In this version the gold was buried in seven cans at the mouth of the Little Colorado River. Bass and Brown set out to look for it, but they had one brush with disaster when their pack animals panicked and tumbled off the trail, and another when, during the return trip, Bass nearly succumbed to heat exhaustion while crossing the Painted Desert. Bass went to southern California to recuperate. Brown and another man continued the search and ungraciously informed Bass that he was cut off from all of the profits.

George McCormick mined in Palisades Canyon from 1893 to 1895. He found the tunnel which had been abandoned by Tanner and French, and agreed that it might indeed be Lee's lost mine. He hired Harry McDonald and another man to help extend the tunnel four hundred feet into the wall. They found no valuable minerals, but drove another tunnel on the other side of the wall before they abandoned the effort.

The long, imaginative tentacles of legend seem to have wound around even the work done by the Stanton railroad survey of 1889-1990. When Robert B. Stanton arrived in San Diego at the conclusion of the monumental Colorado River survey, the San Diego Union of May 3, 1890 (p. 6), reported an extraordinary observation in a quote credited to Stanton: "In the Grand Canyon are some of the richest gold mines ever discovered. I noticed a good many prospectors on the trip, and while at Yuma, on my way here, I was told that three new mining towns have been started near the mouth of the Little Colorado River." Stanton may have been extrapolating from the Lee legend. Perhaps, too, he was just keeping the railroad's interests in the public eye.

The Most Wonderful Gold Mine in the World

In an bizarre story printed in the April 11, 1899 *Denver Investor*, an anonymous "correspondent" reported hearing a tale from four torn and bedraggled men who wandered into a Williams, Arizona, hostelry. M.C. Sharpneck of Council Bluffs, Iowa; Frank C. Randall of Kansas City; and Frank

Lee and W.F. Hull (the same William Hull who had dealings in the Bridal Veil and Hance asbestos mines) of Coconino County, Arizona, had gone in search of the huge copper deposit "years ago discovered by [Hull] in one of his numerous solitary explorations."

Sharpneck had built to order a fourteen-foot canvas-covered boat that could be broken down and carried in pieces. The four took it down the Bright Angel Trail and assembled it at the Colorado River. This would have put them in the area of Bright Angel Canyon and today's heavily visited area around Phantom Ranch. The article indicated that only two men at a time could cross the river in the flimsy boat, so three round trips were made.

When we had all landed safely at the big copper ledge which was the real object of our trip, I was tickled to see the surprise of Sharpneck and Randall. I had only promised them a ledge 100 feed [sic] in width, as I did not wish to be accused of exaggeration. . . . But we found three separate veins of copper, each nearly 100 feet wide, that started at the summit of the granite cliffs, some 800 or 1,000 feet above the river, and converged like the sticks of a fan as they descended the steep pitch, until they joined in one tremendous dyke between 500 and 600 feet across and cropping [out] in some places as high as a man's head. Why, there are thousands of tons of ore in sight—enough to keep a 100-ton smelter busy for a generation! And with unlimited power close by, to move dynamos, it is no difficult problem to raise any number of tons of ore to the summit of the plateau.

The allegation of the giant mineralized "dyke" is almost certainly an elaborated account borrowed from Robert B. Stanton. During Stanton's railroad survey in 1890, he had written in his journal after passing Bright Angel Creek (Smith and Crampton, 1987:173):

We passed yesterday evening the great vein that Jack Sumner and Powell said so much about. It is about 3/4 mile above Bright Angel on the right side. I did not stop and test it for various reasons best known to myself, but located it for future use. From the general appearance I think this is a very rich mineral section.

Here again is circumstantial evidence for some overexuberance on Stanton's part, calculated to underwrite interest in the railroad through the Colorado's canyons.

In Hull's fanciful account, he saw in the schist "garnets of all sizes are sticking as thick as plums in a pudding." He continued, "There are thousands of them there, and some are the size of a man's fist. I call them garnets, but it is by no means certain that they are not some more valuable stone." They were harder than "ordinary" garnets, exceeded only by diamonds.

Sharpneck and Hull were thrown out of the boat on the return trip, and they pulled themselves out a few hundred yards downstream. The next day, "we"—it is not clear whether it was Hull and Sharpneck or all four men—struck out downstream in a horrifying and death-defying ride through rapids and monstrous whirlpools. Then, after a close call worthy of one of Hercules' clashes with guardian titans,

. . . we caught a glimpse of a wonderful sight—a sight so wonderful that it seems like a dream to me, even now. In the somber, half light of those overhanging rock masses gleamed a broad belt of brilliant white quartz—ten or twelve feet wide, it seemed to be—and its whole surface was sparkling and scintillating with great seams and coruscations of yellow gold. It was only a momentary glimpse, for we swept by with the speed of a railroad train; but in that moment I believe we saw more gold than anyone has ever looked at in the same space of time.

When we recovered command of the boat and ourselves—some 500 yards below—we tried to make a landing, but did not succeed. The cliffs there drop sheer into the water, and there is not a spot where one could get a foothold. That gold ledge will have to be reached by a shaft from above, and at best it will be a task requiring ingenuity and money. But where there is gold, capital will always find a way to reach it.

We also saw another great deposit of copper, during that swift ride. . . . We passed a large boulder of quartz, colored green and blue with copper carbonates. It stood by itself, just above water line, and undoubtedly the mother ledge from which it came, was not far off; for such a mass does not travel far. It was cube-shaped, and fully six feet high. We had time to notice, too, as we slid by, that its surface was blotched with great patches of metal; probably sulfate of copper and copper glance.

The truth is, there is no end to the wonders to be seen in the Grand Canyon.

Minerals from Outer Space

Somewhere out in the Canyon are known to be the fragments of one, perhaps two, meteorites. Whatever survived the fiery dash through the earth's atmosphere certainly shattered on impact, and the broken pieces of meteorites that may have been no larger than a person's fist now lie scattered amidst the countless other pieces of rubble from Grand Canyon's majestic walls.

The Hualapai once occupied a village in Meriwitica Canyon, three miles upstream from the mouth of Spencer Canyon. Huntoon (1989c) has recounted:

In 1921 or 1924(?), a meteor streaking through the early evening darkness slammed into the north wall of Milkweed Canyon across from the village (Grant Tapije, personal recollection, 1986). The meteorite impact left a visible but not particularly obvious scar on the Redwall cliff (Young, 1978, p. 288).

Bill Bass, born in 1900, son of Grand Canyon pioneer William Wallace Bass, recounted what was probably a meteorite impact in the Canyon. He was a lad growing up at Bass Camp, when

One evening during the summertime, it was about dusk and I was sitting out there when a bright light appeared, like a falling star, right over the river to the west, halfway across the Canyon. It lit up the whole sky, and after it hit there was a loud noise like an explosion, and for a long time I heard rock rolling . . . (Maurer, 1983:33).

Surely over the millennia, many objects streaked from space the extra mile into the depths of the Canyon. Whether they were stony meteorites, composed of igneous minerals from the shattered remains of moonlets, or iron-nickel masses that once formed the cores of ancient asteroids, we do not know. Nor do we know how many lie out there, unfound, scattered among the rocks on talus slopes and along dry washes.

Admittedly, there is no element in a meteorite that cannot be found on the earth. Still, its remote beginnings in the deep reaches of time and space, and its unimaginably long wandering toward a brilliant end in Earth's atmosphere give us pause for thought and wonder—far more than any deposit of rich metal or mineral can provide. Its incandescent streak to the ground, the sharp crack of impact, rouse even the sleeping mountains from their dreams.

REFERENCES

Anonymous. 1957. "A Bounty from Ancient Bats." *Life*, (November 11):103-104, 106.

___. 1958. "It Comes from Here . . . . and Ends Up Here." *Good Packaging*, (November):14-15.

Ageton, R.W. 1952. Access Road: Copper House Mining Co., Mohave County, Arizona, Engineering Report, 4 pp.

Akens, Jean. 1991. "Ed Smart, Miner." *Canyon Legacy*, (10):14-18.

Allen, M.A., and Butler, G.M. 1921. "Asbestos; Mines and Mineral Resources." University of Arizona Bulletin 113(24), Bureau of Mines and Mineral Technology Series, pp. 21-26.

Altschul, Jeffrey H., and Fairley, Helen C. 1989. "Man, Models and Management: An Overview of the Archaeology of the Arizona Strip and the Management of Its Cultural Resources." Report prepared for U.S. Forest Service and U.S. Bureau of Land Management, contract no. 53-8371-6-0054, submitted by Statistical Research, Plateau Archaeology, Dames and Moore, Inc., 410 pp.

Arizona Champion. 1883. Hackberry, general items of interest about town. Flagstaff, (December 22).

___. 1886. "Fatal Duel—The Cattlemen, Charles Spencer and Charles Cohen Fight to the Death." Flagstaff, (December 4).

Arizona Daily Star. 1884. Tucson, Arizona, 4(3) (June 16).

Arizona Journal-Miner Weekly. 1886. "A Fatal Quarrel." Prescott, 24(126) (November 26).

___. 1886. "The Cohen Examination." Prescott, 24(134) (December 7).

Avery, Valeen Tippetts. 1981. "Free Running; Charlie Spencer and His Most Remarkable Water Project; Being an Account of Speculative Investment in the Golden West of the 1900s." [Flagstaff:] Flagstaff Corral of Westerners International, 29 pp.

Beaman, E.O. 1874. "The Cañon of the Colorado, and the Moquis Pueblos." *Appleton's Journal*, 11(265):481-484, (266):513-516, (267):545-548, (268):590-593, (269):623-626, (270):641-644, (271):686-689.

Beatty, W.B. 1962. "Geology and Mining Operations in U.S. Guano Cave, Mohave County, Arizona." *Cave Notes*, 4(5):40-41.

Belknap, Buzz, and Evans, Lois Belknap. 1989. *Grand Canyon River Guide*. Evergreen, Colorado: Westwater Books, 96 pp.

Belshaw, Mike, and Peplow, Ed Jr. 1983. Historic Resources Study. Volume 2. "Mines and mining districts in the Lake Mead National Recreation Area, and historic sites within Lake Mead National Recreation Area deemed ineligible for National Register nomination, Lake Mead National Recreation Area Arizona-Nevada." Denver, Colorado: Historic Preservation Branch, Pacific Northwest/Western Team, Denver Service Center, National Park Service, U.S. Department of the Interior, Report D-2067, 163 pp.

Beus, Stanley S., and Morales, Michael (eds.). 1990. *Grand Canyon Geology*. New York: Oxford University Press, and Flagstaff: Museum of Northern Arizona Press, 518 pp.

Billingsley, George H. 1974. "Mining in Grand Canyon." In Breed, W.J., and Roat, E.C. (eds.). *Geology of the Grand Canyon*. [Flagstaff:] Museum of Northern Arizona, and Grand Canyon Natural History Association, pp. 170-178.

___. 1986. "Erosional patterns prior to deposition of the Surprise Canyon Formation in the Grand Canyon, Arizona." Geological Society of America Abstracts with Programs, Rocky Mountain Section Meeting, Flagstaff, Arizona, 18(5):341.

Billingsley, G.H., and Elston, D.P. 1989. "Geologic Log of the Colorado River from Lees Ferry to Temple Bar, Lake Mead, Arizona." *In* Elston, D.P., Billingsley, G.H., and Young, R.A. (eds.). "Geology of the Grand Canyon, northern Arizona" (with Colorado River guides). 28th International Geological Congress Field Trip Guidebook T115/315. Washington, D.C.: American Geophysical Union, pp. 1-36.

Billingsley, G.H., and Hendricks, J.D. 1989. "Physiographic Features of Northwestern Arizona." *In* Elston, D.P., Billingsley, G.H., and Young, R.A. (eds.). "Geology of Grand Canyon, northern Arizona" (with Colorado River guides). 28th International Geological Congress Field Trip Guidebook T115/315. Washington, D.C.: American Geophysical Union, pp. 67-72.

Billingsley, G.H., Antweiler, J.C., Beard, L.S., Lucchitta, Ivo, and Lane, M.E. 1986a. "Mineral Resource Potential of the Pigeon Canyon, Nevershine Mesa, and Snap Point Wilderness Study Areas, Mohave County, Arizona." U.S. Geological Survey Miscellaneous Field Studies Map MF-1860-A, 1 sheet, text 9 pp.

Billingsley, G.H., Hendricks, J.D., and Lucchitta, Ivo. 1986b. "Field Guide to the Lower Grand Canyon, from Peach Springs to Pierce Ferry, Arizona." *In* Davis, G.H., and VandenDolder, E.M. (eds.). "Geologic Diversity of Arizona and Its Margins; Excursions of Choice Areas"; Geological Society of America Field Trip Guidebook, 100th Annual Meeting. Arizona Bureau of Geology and Mineral Technology, [Geological Survey Branch] Special Paper 5, pp. 20-38.

Birdseye, R.W. 1925. "Greatest Ladder in the World Built on the Cliff in Havasu Canyon." *Travel*, (46):33.

Bliss, James D. 1993. "Mineral Resource Assessment of Undiscovered Mineral Deposits for Selected Mineral Deposit Types in the Kaibab National Forest, Arizona." U.S. Geological Survey Open-File Report 93-329, 61 pp.

Bowyer, Ben, and Peterson, W.R. 1959. "Examination of the Hacks Canyon Mine, Mohave County, Arizona" for the November 24, 1958 announcement. U.S. Atomic Energy Commission unpublished report, 8 pp. [Cited by Chenoweth (1988).]

Bromfield, C.S. 1952. "Access-road copper: Copper House Mining Co., Mohave County, Arizona," AR Docket 126, Geology, 5 pp.

Brown, J.F. 1955. "Certification of the Orphan Lode Claim, Arizona District, Coconino County, Arizona" (Application

No. 705). U.S. Atomic Energy Commission Report C-731, 6 pp.

Brundy, C.M. 1977. "Orphan with a Midas Touch." *The Denver Post* (Empire Magazine Section), November 27, 1977, pp. 12-17.

Busch, J.E., and Ferris, H.A. 1923. "Havasupai Mining Claims; General Land Office, Santa Fe, New Mexico" from a manuscript copy by Edwin D. McKee at the Museum of Northern Arizona, Flagstaff, Arizona, 7 pp.

___. 1924. "Bridal Veil Mines in Cataract Canyon." *Arizona Mining Journal,* 8:9-10, 23-24.

Butchart, J. Harvey. 1970. *Grand Canyon Treks; A Guide to the Inner Canyon Routes.* Glendale, California: La Siesta Press, 72 pp.

Canty, J. Michael, and Greeley, Michael N. (eds.). 1987. *History of Mining in Arizona.* Tucson: Mining Club of the Southwest Foundation, 279 pp.

___. 1991. (eds.). *History of Mining in Arizona.* Volume II. Mining Club of the Southwest Foundation, Tucson, and American Institute of Mining Engineers, Tucson Section, 293 pp.

Carlson, Helen S. 1974. *Nevada Place Names; A Geographical Dictionary.* Reno: University of Nevada Press.

Chapman, H.H. 1917. "Mining Claims in the Grand Canyon." *American Forestry,* 23(280):225-227.

Chenoweth, William L. 1986. "The Orphan Lode Mine, Grand Canyon, Arizona, A Case History of a Mineralized, Collapse-breccia Pipe." U.S. Geological Survey Open-File Report 86-510, 126 pp.

_____.1988. "The Production History and Geology of the Hacks, Ridenour, Riverview and Chapel Breccia Pipes, Northwestern Arizona." U.S. Geological Survey Open-File Report 88-648, 60 pp.

Chenoweth, William L., and Akers, J.J. 1958. "Road Log from Gray Mountain to the Gap and thence to Desert View—Guidebook of the Black Mesa Basin." New Mexico Geological Society, Ninth Field Conference, pp. 45-53.

Cleland R.G., and Brooks, Juanita. 1983. A *Mormon Chronicle; The Diaries of John D. Lee, 1848-1876*: edited and annotated by Robert G. Cleland and Juanita Brooks. Salt Lake City: University of Utah Press, vol. 1, 344 pp., vol. 2, 480 pp.

Coconino Sun. 1901. "Killed by Falling Rocks." Flagstaff (February 23).

Coder, Christopher M. 1994. "Historical Archaeology." *In*: Fairley, Helen , Bungart, Peter W., Coder, Christopher M., Huffman, Jim, Samples, Terry L., and Balsom, Janet R., "The Grand Canyon River Corridor Survey Project: archaeological survey along the Colorado River between Glen Canyon Dam and Separation Canyon." Submitted by Janet R. Balsom, Principal Investigator. Prepared in cooperation with the Glen Canyon Environmental Studies, Cooperative Agreement No. 9AA-40-07920, pp. 113-146.

Colton, Harold S. 1940. "Tracing the Lost Mines of the Padres." Plateau, 13(2):17-22.

Collins, C.E. 1951. "Certification of the Hacks Mining Company, Fredonia, Arizona." U.S. Atomic Energy Commission Report C-38, 3 pp.

Corle, Edwin. 1946. Listen, Bright Angel. New York: Duell, Sloan and Pearce, 312 pp.

Cox, N.I. 1982. A Harsh Land and Proud, Saga of the Arizona Strip. Las Vegas, Nevada: Cox Printing Co., 386 pp.

Cox, N.I., and Russell, H.B. 1973. Footprints on the Arizona Strip. Bountiful, Utah: Horizon Publishers, 242 pp.

Crampton, C.G. 1972. *Land of Living Rock;* The Grand Canyon and the High Plateaus, Arizona, Utah, and Nevada. New York: Alfred A. Knopf, 268 pp.

Crumbo, Kim. 1995. "Draft environmental assessment; Bat Cave restoration, Grand Canyon National Park." Grand Canyon National Park, 7 pp.

Daily Arizona Miner. 1880. "The Deepest Canyons in the World." Prescott, Arizona, 7(70) (April 7).

___. 1880. "Little Colorado River, Yavapai County." Prescott, Arizona, 7(73) (April 12).

Dawdy, Doris Ostrander. 1993. *George Montague Wheeler; The Man and the Myth.* Athens, Ohio: Swallow Press/Ohio University Press, 122 pp.

Day, Allen. 1905. "Mining in the Grand Canyon." *Scientific American,* 93:478.

Dellenbaugh, Frederick S. 1934. *Indian Red Paint.* Masterkey, 8(3):85.

Denver Investor. 1899. "Great Veins of Ore." (April 11).

Denver Republican. 1892. "Up from Colorado CaÒon, Prospector Beamer Talks of the Chances for Good Ore." (July 17):7.

Dewald, Bud. 1958. "Canyon Cable to Riches; A Two-mile Tram Hauls Ancient Guano from Grand Canyon Cave." *Arizona Republic (Arizona Days and Ways Magazine),* January, p. 12.

Dobyns Henry F., and Euler Robert C. 1976. *The Walapai People.* Phoenix: Indian Tribal Series, 106 pp.

Dodge, Matt, and McKleveen, J.W. 1970. "Hogan's Orphan Mine." *True West,* (December):6-10, 40, 42.

DuBois, S.M., Smith, A.W., Nye N.K., and Nowack, T.A., and Thaddeus A. Jr. 1982. "Arizona Earthquakes, 1776-1980." Arizona Bureau of Geology and Mineral Technology Bulletin 193, 456 pp.

Dunning, C.H. 1948. "A Report on Hack's Canyon Uranium Mine, Mohave County, Arizona." U.S. Atomic Energy Commission Open-File Report RMO-999, 11 pp.

Dunning, C.H., and Peplow, E.H. Jr. 1959. *Rocks to Riches.* Phoenix: South Publishing Co.

Dutton, Clarence E. 1882. *The Tertiary History of the Grand Cañon District; with Atlas.* U.S. Geological Survey Monograph 2, 264 pp. and Atlas.

Elsing, M.J., and Heineman, R.E.S. 1936. "Arizona Metal Production." Arizona Bureau of Mines Bulletin 140, 112 pp.

Elston, Donald P., and Young, Richard A. 1989. "Development of Cenozoic Landscape of Central and Northern Arizona; Cutting of Grand Canyon." *In* Elston, D.P., Billingsley, G.H., and Young, R.A. (eds.). "Geology of Grand Canyon, Northern Arizona (with Colorado River guides)." 28th International Geological Congress Field Trip Guidebook T115/315. Washington, D.C.: American Geophysical Union, pp. 145-155.

___. 1991. "Cretaceous-Eocene (Laramide) Landscape Development and Oligocene-Pliocene Drainage Reorganization of Transition Zone and Colorado Plateau, Arizona." *Journal of Geophysical Research, Series B,* 96(B7):12,389-12,406.

Elston, Donald P., Billingsley, George H., and Young, Richard A. (eds.). 1989. "Geology of Grand Canyon, Northern Arizona (with Colorado River Guides), Lees Ferry to Pierce Ferry, Arizona." 28th International Geological Congress Field Trip Guidebook T115/315, American Geophysical Union, Washington, D.C., 239 p.

Euler, Robert C. 1978. "Archaeological and Paleobiological Studies at Stanton's Cave, Grand Canyon National Park, Arizona—A Report of Progress." National Geographic Society Research Reports, 1969 Projects, pp. 141-162.

Fairley, Helen C., Bungart, Peter W., Coder, Christopher M., Huffman, Jim, Samples, Terry L., and Balsom, Janet R. 1994. "The Grand Canyon River Corridor Survey Project: Archaeological Survey along the Colorado River between Glen Canyon Dam and Separation Canyon." Submitted by Janet R. Balsom, Principal Investigator. Prepared in cooperation with the Glen Canyon Environmental Studies, Cooperative Agreement No. 9AA-40-07920, 276 pp.

Farnham, L.L., and Stewart, L.A. 1958. "Manganese Deposits of Western Arizona." U.S. Bureau of Mines Information Circular 7843.

Fleischner, Michael. 1987. *Glossary of Mineral Species.* Tucson: The Mineralogical Record. 234 pp.

Foord, Eugene E., McKee, Edwin D., and Bowles, C. Gil. [1978]. "Status of Mineral Resource Information for the Shivwits Plateau, Parashant, Andrus, and Whitmore Canyons, and Kanab Canyon Areas, Grand Canyon, Arizona." Administrative report prepared by the U.S. Geological Survey for the National Park Service. 30 pp.

Fukui, L.M. 1982. "Selected Thin-section Analyses." *In* Ballieul, T.A., and Zollinger, R.C., "National Uranium Resource Evaluation, Grand Canyon Quadrangle Arizona." U.S. Department of Energy Preliminary Folio PGJ/F-020(82), Appendix E, 59 pp.

Garbani, James H. 1993. *Arizona Mines and Mining Companies, 1854-1954.* Tucson: Arizona Territorial Trader, 307 pp.

Gibson, Russell. 1952. "Reconnaissance of Some Red Bed Copper Deposits in the Southwestern United States." U.S. Atomic Energy Commission Report RMO-890, 78 pp.

Goodman, David M. 1969. *Arizona Odyssey; Bibliographic Adventures in Nineteenth Century Magazines.* Tempe: Arizona Historical Foundation, 360 pp.

Gornitz, Vivien M. 1969. "Mineralization, Alteration and Mechanism of Emplacement, Orphan Ore Deposit, Grand Canyon, Arizona." Ph.D. dissertation, Columbia University, New York, New York, 186 pp.

Gornitz, Vivien, and Kerr, P.F. 1970. "Uranium Mineralization and Alteration, Orphan Mine, Grand Canyon, Arizona." *Economic Geology*, 65(7):751-768.

Granger, B.H. 1983. *Arizona's Names.* Falconer Publishing Co., distributed by Treasure Chest Publications, Tucson, Arizona, 824 pp.

Grattan, V.L. 1992. *Mary Colter, Builder upon the Red Earth.* Grand Canyon Natural History Association, Grand Canyon, Arizona, 131 pp.

Hall, S.M. 1975. *Sharlot Hall on the Arizona Strip; A Diary of a Journey through Northern Arizona in 1911* (Crampton, Gregory, ed.). Flagstaff, Arizona: Northland Press, 97 pp.

Harrington, M.R. 1925. "Ancient Salt Mine near St. Thomas, Nevada." *Indian Notes*, 2:227-231.

___. 1926. "Another Ancient Salt Mine in Nevada." *Indian Notes*, 3:221-232.

Harris, Grant B. 1980. *Shanley, Pennies Wise—Dollars Foolish.* New York, Washington, Atlanta, Los Angles, and Chicago: Vantage Press.

Hately, J.G. 1907. "Copper Mining on the Colorado River." *Mineral World*, 26:809.

Hausbrand, O., Hoffmann, A., Hülsemann, P., Isert, F., Landschütz, H., von zer Mühlen, L., and Stahl, A. 1927. *Weltlagerstättenkarte / Carte des Gites Minéraux de Monde /* Map of the Mineral Deposits of the World / *Mapa de los Yacimientos del Mundo.* Berlin: Dietrich Reimer/Ernst Vohsen, 8 sheets.

Heikes, V.C. 1919. "Gold, Silver, Copper, Lead, and Zinc in Arizona (mines report)." *In* Mineral resources in the United States, 1916, Part 1, Metals. U.S. Geological Survey, pp. 283-319.

Herefore, Richard. 1993. "Map Showing Surficial Geology and Geomorphology of the Palisades Creek Archeologic Area, Grand Canyon National Park, Arizona." U.S. Geological Survey Open-File Report 93-553, 1 sheet, text 20 pp.

Hewett, D.F., Callaghan, Eugene, Moore, B.N., Nolan, T.B., Rubey, W.W., and Schaller, W.T. 1936. "Mineral Resources of the Region around Boulder Dam." U.S. Geological Survey Bulletin 871, 197 pp.

Hill, J.M. 1915. "The Grand Gulch Mining Region, Mohave County, Arizona." U.S. Geological Survey Bulletin 580, pp. 39-58.

Holt, Albert Bushnell, and Ferleger, Herbert Ronald. 1941. *Theodore Roosevelt Cyclopedia.* New York: Theodore Roosevelt Memorial, 67 pp.

Hughes J. Donald 1967. *The Story of Man at Grand Canyon.* Grand Canyon Natural History Association Bulletin 14, 195 pp.

___. 1978. *In the House of Stone and Light; A Human History of the Grand Canyon.* Grand Canyon Natural History Association, 137 pp.

Huntoon, Peter W. 1970. "The Hydro-mechanics of the Ground Water System in the Portion of the Kaibab Plateau, Arizona." Ph.D. dissertation, University of Arizona, Tucson, 251 pp.

___. 1989a. "Phanerozoic Tectonism, Grand Canyon, Arizona." *In* Elston, D.P., Billingsley, G.H., and Young, R.A. (eds.). Geology of Grand Canyon, Northern Arizona (with Colorado River guides)." 28th International Geological Congress Field Trip Guidebook T115/315. Washington, D.C.: American Geophysical Union, pp. 76-89.

___. 1989b. "Bat Cave Guano Mine, Western Grand Canyon, Arizona." In Elston, D.P., Billingsley, G.H., and Young, R.A. (eds.). "Geology of Grand Canyon, Northern Arizona (with Colorado River guides)." 28th International Geological Congress Field Trip Guidebook T115/315. Washington, D.C.: American Geophysical Union, p. 228.

___. 1989c. "Small Meteorite Impact in the Western Grand

Canyon, Arizona." *In* Elston, D.P., Billingsley, G.H., and Young, R.A. (eds.). "Geology of Grand Canyon, Northern Arizona (with Colorado River guides)." 28th International Geological Congress Field Trip Guidebook T115/315. Washington, D.C.: American Geophysical Union, p. 228.

Ives, J.C. 1859. "The Colorado Expedition." *Journal of the American Geographical and Statistical Society*, 1:41-45.

James, George Wharton. 1901. *In and Around the Grand Canyon; The Grand Canyon of the Colorado River in Arizona*. Boston: Little, Brown and Co., 352 pp.

Jennings, E.P. 1904. "The Copper Deposits of the Kaibab Plateau, Arizona." Transactions of the American Institute of Mining Engineers, 24:839-841.

Johnson, W.I. [No date]. Northern Arizona Lead and Zinc Mining Corporation [prospectus]. Prescott, Arizona, 8 pp.

Kane County Independent. 1912. "Mining Boom Busted; Pahreah and Lees Ferry Co's in Financial Trouble." (April 11): 1.

Kiesel, H.C. 1873. "Exploring Expedition." *Salt Lake Tribune*, (July 30).

Kildare, Maurice. 1975. "John D. Lee's Secret Mine." *In Treasure trails of the Old West*. True Treasure Publications. pp. 17-23.

Kofford, M.E. 1969. "The Orphan Mine." *In* Baars, D.L., ed., Four Corners Geological Society, 5th Field Conference, Guidebook, pp. 190-194.

Kolb, Ellsworth L. 1914. *Through the Grand Canyon from Wyoming to Mexico*. New York: The MacMillan Company, 344 pp. [Many reprintings, with variations.]

Kolb, Ellsworth L., and Kolb, Emery. 1914. "Experiences in the Grand Canyon." *National Geographic*, 26(2):99-184.

Kroeber, A.L. 1935. (Ed.) "Walapai Ethnography." Memoirs of the American Anthropological Association, (42).

Lavender, David. 1985. *River Runners of the Grand Canyon*. Grand Canyon Natural History Association, 188 pp.

Lane, M.E. 1984. "Mineral Investigations of the Pigeon Canyon, Nevershine Mesa, Snap Point, and Last Chance Wilderness Study Areas (BLM), Mohave County, Arizona." U.S. Bureau of Mines Open-file Report MLA 84-8, 61 pp.

Larson, Andrew Karl. 1979. *I Was Called to Dixie; The Virgin River Basin: Unique Experiences in Mormon Pioneering*. [No imprint; privately published.]

Lauzon, H.R. 1934. "Is There Gold in the Canyon?" *Grand Canyon Nature Notes*, 8(10):233-234.

Leicht, W.C. 1971. "Minerals of the Grandview Mine." *Mineralogical Record*, (September-October):215-221.

Lenz, Lee W. 1986. *Marcus E. Jones; Western Geologist, Mining Engineer and Botanist*. Claremont, California: Rancho Santa Ana Botanic Garden, 486 pp.

Lockwood, F.C. 1940. "Captain John Hance." Desert, 3(9):15-18.

___. 1942. "More Arizona Characters." *University of Arizona General Bulletin*, 13(3):41-52.

Logan, Roger V. Jr. 1992. "New Light on the Mountain Meadows Caravan." *Utah Historical Quarterly*, 60(3):224-237.

Longwell, Chester R. 1946. "How Old Is the Colorado River?" *American Journal of Science*, 244:817-835.

Lovejoy, E.M.P. 1954. "Results of an Aerial Radiometric

Examination of the Ridenour Mine District, Hualapai Indian Reservation Coconino County, Arizona." *In* Supplement to the U.S. Atomic Energy Commission, Report RME 2014, Part 1, pp. 19-22.

Lucchitta, Ivo. 1984. "Development of the Landscape in Northern Arizona—The Country of Plateaus and Canyons." In Smiley, T.L., Nations, J.D., Péwé, T.L., and Schafer, J.P. (eds.). *Landscapes of Arizona*. New York: University Press of America, pp. 269-302.

___. 1988. "Canyon Maker; A Geological History of the Colorado River." *Plateau*, 39(2), 32 pp.

Ludwig, K.R., and Simmons, K.R. 1992. "U-Pb Dating of Uranium Deposits in Collapse Breccia Pipes of the Grand Canyon Region." *Economic Geology*, 87:1747-1765.

Malach, Roman. 1975. *The Arizona Strip in Mohave County*. Kingman, Arizona: Mohave Pioneers Historical Society, Arizona Bicentennial Commission; printed by Graphicopy, New York, 48 pp.

Mary, R.F., and Colton, H.S. 1931. Petroglyphs; The Record of a Great Adventure." *American Anthropologist*, 33(1):32-37.

Maurer, Stephen G. 1983. (Based on conversations with William G. Bass) *Solitude and Sunshine; Images of a Grand Canyon Childhood*. Boulder, Colorado: Pruett Publishing Co., 97 pp.

Mazzu, Linda , with Rihs, John. [No date.] "Final Report; Intensive Reconnaissance Sampling of Grand Canyon Tributaries, Grand Canyon National Park." Report submitted to National Park Service, Grand Canyon National Park, [17] pp. [Late 1994 or early 1995.]

McKee, Edwin D. 1933. "On Canyon Trails." *Grand Canyon Nature Notes*, 8(6):191-194.

___. 1982. "The Supai Group of Grand Canyon." U.S. Geological Survey Professional Paper 1173, 504 pp.

Miller, R.D. 1954a. "Reconnaissance for Uranium in the Hualapai Indian Reservation Area, Mohave and Coconino Counties, Arizona." U.S. Atomic Energy Commission Report RME-2007, 18 pp.

___. 1954b. "Copper-uranium Deposit at the Ridenour Mine, Hualapai Indian Reservation, Coconino County, Arizona." Oak Ridge, Tennessee: U.S. Atomic Energy Commission, Technical Information Service, Part 1, pp. 1-18.

Mining World. 1959. "How Western Gold Mines Uranium in Grand Canyon." 1(1):32-35.

Mohave County Miner. 1904. [Item including mention of platinum ore.] (April 16).

___. 1914. "Grand Gulch Mines." (July 25), pp. 1, 6.

Mohave Magazine. 1973. Supplement to Lake Havasu Herald, Mohave County Miner, and *Mohave Valley News*, 2(1) (April 5).

Murbarger, Nell. 1958. "Trail Blazer of Grand Canyon." *Desert*, (21):5-9.

National Park Service. 1962. "Orphan Mining Claim in Grand Canyon National Park, Arizona, Deeded to the Federal Government." U.S. Department of the Interior News Release 4895. [Museum of Northern Arizona separates, no. 11,059.]

___. 1977. "Environmental Assessment, Management Option for

Examination of Uranium Leases, Lake Mead National Recreation Area, Arizona-Nevada." National Park Service, Denver Service Center, Report D-1190, 123 pp.

National Wildlife News. 1962. "House Rejects Grand Canyon Mining Claim Bill in Hot Debate." 3(4).

___. 1962. "Becomes Law." 3(7).

Nelson, H.E., and Rambosek, A.J. 1970. "Preliminary Reconnaissance Report, Chapel Claim." *In* Grand Junction Office, Atomic Energy Commission, and U.S. Geological Survey, Preliminary reconnaissance for uranium in Mohave County, Arizona, 1952 to 1956. U.S. Atomic Energy Commission Report RME-158, pp. 148-151.

Nelson, H.E., and Steinhauser, S.R. 1952. "Report on the Examination of the Hacks Canyon Mine, Mohave County, Arizona." U.S. Atomic Energy Commission unpublished report, 8 pp. [Cited by Chenoweth (1988).]

New York Times. 1957. "Helicopter Builds an Aerial Span to Guano Cave in Grand Canyon." (March 27):43, 47.

Newberry, J.S. 1861. "Geological Report." *In* Ives, J.C., "Report upon the Colorado River of the West, Explored in 1857 and 1858." 36th Congress, 1st Session, House Executive Document 90, Part 3, 154 pp. [separately paginated section in volume].

Noble, L.F. 1914. "The Shinumo Quadrangle." U.S. Geological Survey Bulletin 549, 100 pp.

Nuclear Exchange Corporation. 1978. "Uranium Occurrences in Breccia Pipes." Nuclear Exchange Corporation (Nuexco) Report 123, pp. 2.0-2.2.

Osterstock, R.W. 1954. "Supplemental Geologic Report on DMEA Docket 3075 (Copper-Uranium); J and M Leasing Co., Copper Mountain Mine, Cox and Ross Claims, Mohave County, Arizona." U.S. Geological Survey unpublished report, 6 pp.

Phalen, W.C. 1919. *Salt Resources of the United States.* U.S. Geological Survey Bulletin 669, 284 pp.

Poyner, D.K. 1972. "Preliminary and Summary Report, Chapel House Prospect (Knight Lease), Shivwits Plateau, Mohave County, Arizona." U.S. Atomic Energy Commission unpublished report, 12 pp.

Pratt, J.H. 1905. "Arizona Asbestos Deposits." *Mineral World,* 23:17.

Price, V.N., and Darby, J.T. 1964. "Preston Nutter, Utah Cattleman, 1886-1936." *Utah Historical Quarterly,* 32:232-251.

Pyne, Stephen J. 1982. *Dutton's Point; An Intellectual History of the Grand Canyon.* Grand Canyon Natural History Association Monograph 5, 64 pp.

Rasor, C.A. 1949. "Report on Investigation of Radiative Minerals at Hack's Canyon Mine, Mohave County, Arizona." U.S. Atomic Energy Commission open-file report RMO-24, 7 pp.

Raymond, Rossiter W. 1877. *Statistics of Mines and Mining in the States and Territories West of the Rocky Mountains; Being the Eighth Annual Report of Rossiter W. Raymond, United States Commissioner of Mining Statistics.* Washington, D.C.: U.S. Government Printing Office, 519 pp.

Reynolds, M.W., Palacas, J.G., and Elston, D.P. 1989. "Potential Petroleum Source Rocks in the Late Proterozoic Chuar Group, Grand Canyon, Arizona." *In* Elston, D.P., Billingsley, G.H., and Young, R.A. (eds.). "Geology of Grand Canyon, Northern

Arizona (with Colorado River guides)." 28th International Geological Congress Field Trip Guidebook T115/315. Washington, D.C.: American Geophysical Union, p. 117.

Richardson, Gladwell. 1953. "Gold behind a Waterfall." *Desert,* 16(3):22-25.

Richmond, Al. 1985. *Cowboys, Miners, Presidents, and Kings; The Story of the Grand Canyon Railway.* Flagstaff, Arizona: Grand Canyon Pioneers Society. 187 pp.

Ringholz, Raye C. 1989. *Uranium Frenzy; Boom and Bust on the Colorado Plateau.* [Albuquerque:] University of New Mexico Press, 310 pp.

Robertson, D.B. 1986. *Encyclopedia of Western Railway History; The Desert States; Arizona, Nevada, New Mexico, Utah.* Caldwell, Idaho: Caxton Printers, 318 pp.

Rusho, W.L., and Crampton, C.G. 1975. *Desert River Crossing; Historic Lees Ferry on the Colorado River.* Salt Lake City, Utah, and Santa Barbara, California: Peregrine Smith. 126 pp.

___. *Lee's Ferry; Desert River Crossing.* Salt Lake City: Cricket Productions, 168 pp.

Salt Lake Tribune. 1879a. "The New Mining District." (January 21):1.

___. 1879b. "The Prospectors." (February 7).

___. 1879c. "The Prospectors." (February 9).

San Diego Union. 1890. "Engineer Stanton." (May 3): 6.

Schmidt, K.H. 1989. "The Significance of Scarp Retreat for Cenozoic Landform Evolution on the Colorado Plateau, U.S.A." *Earth Surface Processes and Landforms,* 14:93-105.

Schrader, F.C. 1909. "Mineral Deposits of the Cerbat Range, Black Mountains and Grand Wash Cliffs, Mohave County, Arizona." U.S. Geological Survey Bulletin 397, pp. 142-150.

Seargent, H.H. 1959. "Mooney Falls, Havasu Canyon; The Story of Its Naming as Well as the Other Falls." *Arizona Highways,* 35(8):18-28.

Smith, D.C., and Crampton, C.G. (eds.). 1987. *The Colorado River Survey; Robert Brewster Stanton and the Denver, Colorado Canyon and Pacific Railroad.* Salt Lake City: Howe Brothers, 305 pp.

Spamer, Earle E. 1983. *Geology of the Grand Canyon; An Annotated Bibliography (1857-1982) with an Annotated Catalogue of Grand Canyon Type Fossils.* Geological Society of America, Microform Publication 13, 544 pp.

___. 1984a. "Paleontology in the Grand Canyon of Arizona; 125 Years of Lessons and Enigmas from the Late Precambrian to the Present." *Mosasaur,* 2:45-128.

___. 1984b. *Geology of the Grand Canyon; An Annotated Bibliography with an Annotated Catalogue of Grand Canyon Type Fossils, Volume 2, Supplement to the bibliography (1857-1982), Supplement and revisions to the catalogue.* Geological Society of America, Microform Publication 14, 232 pp.

___. 1988. *Geology of the Grand Canyon; An Annotated Bibliography with an Annotated Catalogue of Grand Canyon Type Fossils, Volume 3, Second supplement (to 1987), with an annotated catalogue of the world literature on the Grand Canyon type fossil Chuaria circularis Walcott, 1899, an index fossil for the Late Proterozoic.* Geological Society of America Microform Publication 17, 343 pp.

___. 1989. *The Development of Geological Studies in the Grand*

Canyon. Tryonia, (17), 87 p.

___. 1990a. *Geology of the Grand Canyon; An Annotated Bibliography with an Annotated Catalogue of Grand Canyon Type Fossils, Volume 4, Third supplement (to 1989), with supplement to the annotated bibliography of the world literature on the Grand Canyon type fossil Chuaria circularis Walcott, 1899*. Geological Society of America Microform Publication 20, 178 pp.

___. 1990b. *Geology of the Grand Canyon: A Guide and Index to Published Graphic and Tabular Data (excluding paleontology)*. Geological Society of America Microform Publication 21, xlix + 674 pp.

___. 1990c. (with contributions by George H. Billingsley, William J. Breed, Robert C. Euler, Dorothy A. House, Grace Keroher, Valerie Meyer, Richard Quartaroli, and Lawrence E. Stevens) *Bibliography of the Grand Canyon and the Lower Colorado River; from 1540*. Grand Canyon Natural History Association Monograph 8, 12 parts, separately paginated, 370 pp.

___. 1992a. *Geology of the Grand Canyon: An Annotated Bibliography, with an Annotated Catalogue of Grand Canyon Type Fossils. Vol. 5. Fourth supplement (to 1991) with second supplement to the annotated bibliography of the world literature on the Grand Canyon type fossil Chuaria circularis Walcott, 1899, and with author and chronological indexes to Volumes 1-5*. Geological Society of America Microform Publication 23, 234 pp.

___. 1992b. *Geology of the Grand Canyon: A Guide and Index to Published Graphic and Tabular Data (excluding paleontology). Vol. 2*. Geological Society of America Microform Publication 22, 125 pp.

___. 1992c. *The Grand Canyon Fossil Record; A Source Book on Paleontology of the Grand Canyon and Vicinity, Northwestern Arizona and Southeastern Nevada*. Geological Society of America Microform Publication 24, 1,008 pp.

___. 1993a. "Paleontology in the Grand Canyon of Arizona: 125 Years of Lessons and Enigmas from the Late Precambrian to the Present." *Mosasaur*, 2:45-128.

___. 1993b. (with Daniel F. Cassidy and John Irwin) *Bibliography of the Grand Canyon and the Lower Colorado River; from 1540*. Grand Canyon Natural History Association Monograph 8, Supplement 1, separately paginated parts.

Spamer, Earle E., and Bogan, Arthur E. 1993. "Mollusca of the Grand Canyon and Vicinity, Arizona." Proceedings of the Academy of Natural Sciences of Philadelphia, 144:21-68.

Spence, Clark C. 1993. *Mining Engineers and the American West; The Lace-boot Brigade, 1849-1933*. Moscow: University of Idaho Press, 407 pp. [Originally published 1970 by Yale University Press.]

Stanton, Robert Brewster. 1965. *Down the Colorado*: Smith, D.L., ed., Norman, Oklahoma, University of Oklahoma Press, 237 pp.

Stevens, Larry. 1995. *The Colorado River in Grand Canyon; A Comprehensive Guide to Its Natural and Human History*. Flagstaff: Red Lake Books, 4th ed., 116 pp.

Strong, D.H. 1969. "The Man Who Owned Grand Canyon." *American West*, 6(5):33-40.

Sturdevant, Glen E. 1928. "A Visit to an Unfrequented Part of the Grand Canyon." *Grand Canyon Nature Notes*, 2(9):1-2.

Summons, R.E., Brassel, S.C., Eglinton, Geoffrey, Evans, Evan,

Horodyski, Robert J., Robinson, Neil, and Ward, D.M. 1988. "Distinctive Hydrocarbon Biomarkers from Fossiliferous Sediment of the Late Proterozoic Walcott Member, Chuar Group, Grand Canyon, Arizona." *Geochimica et Cosmochimica Acta*, 52:2625-2637.

Swapp, C.W. 1956. "Arizona, Boom or Bust?" *World Oil*, 143(1):81-85.

Tainter, S.L. 1947. "Apex Copper Property, Coconino County, Arizona." U.S. Bureau of Mines Report of Investigations RI-4013, 23 pp.

Tanner, George S. 1974. *John Tanner and His Family*. John Tanner Family Association.

Taylor, Vernon. 1968. [Mining claim found on the Hopi Sipapu, canyon of the Little Colorado River, dated July 4, 1929. Photocopy, Museum of Northern Arizona Special Collection.]

Thybony, Scott. 1980. *A Guide to the Hermit Trail*. Grand Canyon Natural History Association, 32 pp.

___. 1994. *Official Guide to Hiking the Grand Canyon*. Grand Canyon Natural History Association (revised December 1996), 64 pp.

Tillman, C.G. 1954. DMEA Docket 3075 (Copper-Uranium); J and M Leasing Co., Cox and Ross Claims, Mohave County, Arizona. U.S. Geological Survey unpublished report, 8 pp.

Titiev, Misha. 1937. "Hopi Salt Expedition." *American Anthropologist*, 39(2):244-258.

The Union. 1897. "That Arizona Strip." (St. George, Utah), 10(10) (March 6):1.

United States District Court for the District of Arizona, 1960, No. 642-Prescott, in the National Archives, Pacific Southwest Region, Fred L. Baker & Associates, court reporters, 138 pp.

Van Gosen, B.S., Wenrich, K.J., Sutphin, H.B., Scott, J.H., and Balcer, R.A. 1989. "Drilling of a U-mineralized Breccia Pipe near Blue Mountain, Hualapai Indian Reservation, Northern Arizona." U.S. Geological Survey Open-File Report 89-100, 80 pp.

Verbeek E.R., Grout, M.A., and Van Gosen, B.S. 1988. "Structural Evolution of a Grand Canyon Breccia Pipe; The Ridenour Copper-vanadium-uranium Mine, Hualapai Indian Reservation, Coconino County, Arizona." U.S. Geological Survey Open-File Report 88-6, 75 pp.

Verkamp, Margaret M. 1940. "History of Grand Canyon National Park." Master's thesis, University of Arizona, Tucson, Arizona, 70 pp.

___. 1993. *History of Grand Canyon National Park*. Flagstaff, Arizona: Grand Canyon Pioneers Society Collectors Series, Vol. 1, 57 pp. [First publication of Verkamp's (1940) Master's thesis.]

Waesche, H.H. 1933. "Anita Copper Mine." *Grand Canyon Nature Notes*, 7(11):108-112.

___. 1934. "The Grandview Mine Copper Project, History, Geology, and Mining Methods." *Grand Canyon Nature Notes*, 8(12):250-258.

Walcott, Charles D. 1884. "Report of Mr. Charles D. Walcott (Field Work near Eureka, Nevada, and in Eastern Part of Grand Canyon of the Colorado in Arizona)." U.S. Geological Survey, 4th Annual Report, pp. 44-48.

Walker, Dale E. 1975. *Death was the Black Horse; The Story of Rough Rider Buckey O'Neill*. Austin, Texas: Madrone Press. [Reprinted in 1983: *Buckey O'Neill, The Story of a Rough Rider*. University of Arizona Press, Tucson, 200 pp.)

Watkins, T.A. 1976. "The Geology of the Copper House, Copper Mountain, and Parashant Breccia Pipes, Western Grand Canyon, Mohave County, Arizona." Master's thesis, Colorado School of Mines, 91 pp.

Way, T.E. 1980. *A Summary of Travel to Grand Canyon*. Prescott, Arizona: Prescott Graphics, 14 pp.

Weber, Steven A., and Seaman, P. David (eds.). 1985. *Havasupai Habitat*, A.F. Whiting's Ethnography of a Traditional Indian Culture. Tucson: University of Arizona Press, 288 p.

Webb, Roy. 1994. *Call of the Colorado*. Moscow, Idaho: University of Idaho Press, 175 pp.

Weekly Arizona Miner. 1877. "Another Mining Section." Prescott, 1(5) (December 14).

Weekly News, Official Paper of Coconino County. 1897. "A Mining Magnet." Flagstaff, Arizona, 5(36):1 (April 17).

___. 1901. "Death of a Pioneer." Flagstaff, Arizona (February 23).

Wenrich, Karen J. 1988. "Recognition of Breccia Pipes in Northern Arizona." Fieldnotes (Arizona Bureau of Geology, Minerals, and Technology), 18(1):1-5, 11.

Wenrich, Karen J., and Billingsley, George H. 1986. "Field Trip Logƒ Breccia Pipes in Northern Arizona." *In*: Nations, J. Dale, Conway, Clay M., and Swann, Gordon A. (eds.). "Geology of Central and Northern Arizona; Field Trip Guidebook for Geological Society of America Rocky Mountain Section Meeting, Flagstaff, Arizona, 1986," pp. 43-58.

Wenrich, Karen J., and Hlava, Paul F. 1993. "Nickel-cobalt-iron-copper Sulfides and Arsenides in Solution-collapse Breccia Pipes, Northwestern Arizona" [abstract]. *Geological Society of America Abstracts with Programs*, 25(5):162.

Wenrich, Karen J., and Sutphin, Hoyt B. 1987. "Unique Minerals from Redwall Limestone Caves, Arizona; Their Association with Mineralized Breccia Pipes." Geological Society of America Abstracts with Programs, 19(6):463.

Wenrich, Karen J., Billingsley, George H., and Huntoon, Peter W. 1986. "Breccia Pipe and Geologic Map of the Northwestern Hualapai Indian Reservation and Vicinity, Arizona." U.S. Geological Survey Open-File Report 86-458A, 1 sheet, scale 1:48,000, text 29 pp.

Wenrich, Karen J., Billingsley, George H., and Van Gosen, B.S. 1989a. "The Potential of Breccia Pipes in the National Tank Area, Hualapai Indian Reservation, Arizona." U.S. Geological Survey Bulletin 1683-B, 34 pp.

Wenrich, Karen J., Sutphin, Hoyt B., Van Gosen, B.S., and Billingsley, George H. 1989b. "Exploration for Music Mountain Type Quartz-vein Gold Deposits of the Hualapai Indian Reservation, Arizona." 28th International Geological Congress, Washington, D.C., Abstracts, p. 3-349.

Wheeler, George M. 1871. *Preliminary Report Concerning Explorations and Surveys Principally in Nevada and Arizona*. Washington, D.C.: U.S. Government Printing Office, 96 pp. [1872].

White, J.S. Jr. 1987. Appendix. In Fleischer, Michael, Glossary of mineral species. Tucson: The Mineralogical Record. pp. 228-234.

Williams News. 1897. The copper mines. Williams, Arizona, 5(49):1 (July 17).

___. 1897. Down Hogan's Slide. Williams, Arizona (July 24).

___. 1899. Arizona is a grab-bag and Coconino County is one of the best and richest treasures in it. Williams, Arizona, 8(11):1 (October 14).

Wilson, Eldred D. 1962. "A Résumé of Geology of Arizona." Arizona Bureau of Mines Bulletin 171, 140 pp.

Winship, G.P. 1896. "The Coronado Expedition, 1540-1542." U.S. Bureau of American Ethnology, 14th Annual Report, Part 1, pp. 329-613.

Woods, G.K. 1899. (Compiler) *Personal Impressions of the Grand Cañon of the Colorado River*. San Francisco: Whitaker Ray Co., 164 p.

Wyman, Richard. 1956. "Uranium Discovery at Grand Canyon." *Desert*, 19(2):38.

Young, E.J., Weeks, A.D., and Meyerowitz, R. 1966. "Coconinoite, A New Uranium Mineral from Arizona and Utah." American Mineralogist, 51:651-663.

Young, Richard A. 1978. "Discussion of Cambrian Stratigraphic Nomenclature and Ground Water Prospecting Failures on the Hualapai Plateau, Arizona." *Ground Water*, 16:287-289.

___. 1985. "Geomorphic Evolution of the Colorado Plateau Margin in West-central Arizona, A Tectonic Model to Distinguish between the Causes of Rapid, Symmetrical Scarp Retreat and Scarp Dissection." In Hack, J.T., and Morisawa M. (eds.). *Tectonic geomorphology, Binghamton Geomorphology Symposium in Geomorphology International Series*, 15. London: Allen, and Unwin, pp. 261-276.

Young, W.E. 1954a. "Engineering Report DEMA 3075; J and M Leasing Co., Cox and Ross Claims, Mohave County, Arizona." U.S. Geological Survey unpublished report, 19 pp.

___. 1954b. "Supplemental Engineering Report DMEA 3075; J and M Leasing Co., Cox and Ross Claims, Mohave County, Arizona." U.S. Geological Survey unpublished report, 4 pp.

APPENDIX 1

Summary of Mining Activities in the Grand Canyon Region

Pre-Columbian and Historic American Indian Activities
Local sources throughout the Grand Canyon region
Prehistoric/historic - Collected kaolin, salt, gypsum, malachite, red and yellow ocher (hematite), marcasite, calcite, quartz, travertine (aragonite and calcite), chert, fossil crinoid stems. Non-indigenous items obtained in trade: turquoise, obsidian

Spanish Incursionary Period
Vicinity of Red Butte
Mid- to late 1500s - James (1901) indicated that silver was mined, but there is no documentation for the statement.

Bentley Mining District
West of Grand Wash Cliffs
Mid-1800s - Gold placer claims

Grand Gulch Mine (breccia pipe)
Minerals noted: brochantite, azurite, chalcocite, chrysocolla, cotunnite, copper-descloizite, malachite, limonite
1853 June - Copper prospect of Richard Bentley, Samuel S. Adams, and Joseph Cunningham; said to have been sold to Adams and Cunningham by an American Indian
1870 - Adobe smelter built
1873 June 23 - Records reassigned to Mohave County, Arizona
1881 16 May - Sold for delinquent taxes
1882 Feb 3 - Redeemed by C.C. Bradley for $6.97
1886 - Frank Jennings, agent
1890-1892 - "Owners unknown"
1893 - Claimed by Jennings brothers and Adams
1906-1917 - Ore concentrate >14% copper shipped
1906-1958 - Intermittent work
1911 - Richest copper mine in Arizona
1917 or early 1918 - Mine burned
1951 - Total production to date 15,701 tn ore, 24,349 oz silver, 6,651 lb copper, 715 lb lead
1955 - Main buildings burned
1955-1961 - Mine dump worked; drilling and testing
1978 - U.A. Small and C.H. Englehart, owners

Savanic Mine (fault and joint zone at edge of breccia pipe)
Minerals noted: azurite, malachite, chalcopyrite, bornite, chalcocite, hematite, uranium minerals
1878 Apr 20 - Bronze L Mine located by James R. Cunningham
1880 June 13 - Acquired by George H. Dodge et al.; named Savanic
1906 - Controlled by Levi Syphus and Harry Gentry
1906-1919 - $300,000 copper production; no later records of production
1959 - W.E. Covey, owner
1978 - 5M Mining Corporation, owner
1983 Apr - 5M Mining Corporation and Uranez USA planned joint uranium exploration project

Cunningham Mine
Minerals noted: copper and uranium minerals, barite
(?) 1878 Apr 20 - Copper prospect located by James R. Cunningham(?)
1978 - 5M Mining Corporation and Uranez Company referred to scattered diggings as Cunningham Mine

Hidden Canyon prospect (breccia pipe)
Minerals noted: azurite, malachite; iron and uneconomical uranium minerals
1976 - Unidentified party made drillings
1982 Apr 14 - Pathfinder Mines Corporation made claims, exploratory drillings
1982 - Energy Fuels staked claims near mines; not worked

Fort Garrett (basalt dike)
1935 Aug 31 - Claim by W.H. Garrett; claim note found fall 1987

Music Mountain and Lost Basin Mining Districts
Music Mountain Mine
1886 - Gold and copper prospects claimed
1879 or 1880 - Gold discovered along Grand Wash Cliffs; David Southwick et al.
   -Ellen Jane Mine produced gold; also other mines; gold in quartz veins associated with diabase dikes intruding Proterozoic granites and schists
1882 Apr 13 - William B. Ridenour located placer claims, eastern slopes of Music Mountains

Spencer Canyon
After 1921 - Gold placer operation by unknown party; discovered 1988

Bat Cave Mine (bat guano deposit)
1930s - Discovered by a young man on a river trip, later mining attempt by Merle Emery and Beal Masterson; sold to King-Finn Fertilizer Company
1958 - Sold to U.S. Guano Corporation; cable tram constructed
1960 - Guano exhausted; cable clipped by low-flying jet
1961 - Fred Harvey Company approached to buy property for tourist use
1995 - Towers on north side near mine slated for demolition
Present - Hualapai Indian Tribe use cable house on Guano Point for tourist use

Nelson Mine (limestone quarry)
Mid-1890s - Grand Canyon Lime and Cement Company; later Flintkote Limestone Company
Present - Gemstar Lime Company

Hualapai Indian Reservation
   -Anomalous concentrations of gold detected in laboratory analyses of samples from one tributary to Diamond Creek

Centennial Mining District
Unestablished prospects
Ca. early 1880s - Copper mining claims in Andrus Canyon

Snyder Mine (breccia pipe in Watahomigi Formation)
Elements and minerals noted: *chalcocite, azurite, malachite; iron minerals; arsenic, cobalt, nickel, molybdenum, zinc*
1890s - Old Man Snyder and son prospected in western Grand Canyon
1912 Jan - Kolb brothers meet one of the Snyders, Jud or John, near Granite Spring Rapid
1920s - H.W. Roup and W.H. Davies said to have relocated the Snyder Mine
1958 Oct 15 - Last claims by Robert Williforb and Glen Summerlin; named Futile Mine

1975 - Mine relocated by George and Susan Billingsley in 214 Mile Canyon

Old Bonnie Tunnel (breccia pipe in Redwall Limestone)
Elements and minerals noted: *malachite, azurite, arsenic, silver, cobalt, mercury, nickel, molybdenum, lead, zinc*
1890s(?) - Mine located possibly by the Snyders
1940 June 5 - First definite claim, by H. Loraine Cox, Lee Cox, Roy Wood, and Perry Wood; named Copper Flat

Mount Trumbull Mining District
Prospect 20 miles north of Whitmore Canyon (stratiform breccia)
Minerals noted: *malachite and azurite*
1994 - Copper workings found in stratiform breccia; date possibly from 1940s

Copper Mountain Mine (breccia pipe in Esplanade Sandstone)
Elements and minerals noted: *copper, silver, lead, zinc, gold*
1875 Apr 10 - Mine located
1875 May 19 - Incorporated in St. George, Utah
1881 Feb 21 - Reassigned to Mohave County, Arizona, by A.P. Hardy, E.G. Wooley, Robert C. Lund, James Andrus, E.D. Wooley Jr., and E.W. Snow
1890-1914 - Owned by John A. Swapp sometime during this period
Ca. 1915 - Owned by "McMasters," worked by Bishop Whitehead
1941 - Owned by H. Loraine Cox, Lee Cox, and Etta Cox
Late 1940s-early 1950s - Mine intermittently active
Post-World War II - Western Gold prospected for gold and uranium
1955-1961 - Signal Oil and Gas Company intermittently reworked mine dumps
1964 - Claim incorporated within Lake Mead National Recreation Area; unpatented
Mid-1970s - Owned by Exxon Oil Company as a uranium prospect

Chapel House Mine (breccia pipe in Hermit Shale)
Minerals noted: *uranophane, autunite, malachite, azurite, covellite, brochantite, torbernite, pitchblende; unidentified silver and secondary uranium minerals*
1875 Apr 10 - Located by Daniel Seegmiller and E.W. McIntyre; named the Copper Head
1881 Feb 22 - Claim refiled
1952 - Claimed by C.M. Bundy and Omer Bundy
1968 - Cotter Corporation cut drill holes, subleased from Knight
1972 - Claim held by Audrine G. Knight under bonded lease with Bureau of Land Management and National Park Service
1974 Nov 1 - Lease transferred to Exxon Minerals, USA
1978 Dec 9 - Lease renewal denied

Other Claims in Parashant Area
Johnson placer claims
1893 June 4 - Located by J.S. Johnson, W.M. Johnson, H.E. Judd, S.N. Johnson, J.B. Francis, Joe Meeks, C.A. Huntington, Jeremiah Johnson; gold placer claims

Copper claims 6 miles east of Copper Mountain Mine
1880 Nov - Located by B.M. Ellenbeck, C.A. Smith, John H. Cassidy, John Quillan, Mat O. Loughlin, and John H. Rice

"Centennial Mining District"
1883 May - Several copper claims located by W.D. Chidester and J.D.L. Pearce; no worthwhile production

Copper House claim (breccia pipe in Hermit Shale)
Minerals noted: *malachite, azurite, chalcocite, limonite, hematite*

Little Chicken Mine (breccia pipe in Mooney Falls Member of Redwall Limestone)
Elements and minerals noted: *gold (in chemical analysis), zinc, lead, arsenic, barium, cadmium, silver*
date? - Brady Inglestead located copper and silver mine; later claims include Henry Covington
1918 - Mine known as the "Little Chicken"
1929 - Henry Covington refiled claim, together with other claims: Ram, Shepherds Folly, and Golden Slide
1950s - Claims declared null and void by Department of the Interior, but some operations continued
1961 - Road constructed on the Esplanade
1971 - Improvements declared improper; ownership reverted to national park

Brady Canyon claim
1925 - Brady Inglestead, Frank Heaton, and one other running a tunnel to reach water

Tuckup Canyon Mine (apparently in slump block of Toroweap[?] Formation resting on Esplanade Sandstone)
Elements noted: *copper, gold, silver, lead*
Ca. 1911 - First worked by Joe Price and "Hunsucker"
Ca. 1921 - Worked by Marcell Schmutz and father
1936 Sept 5 - Walt and Rasco Cunningham laid claim to mine; named Pinto Mine
1937 Mar 15 - Claim transferred to Wallace Blake
1938 - Claim transferred to William Cumingham; renamed Grand Canyon Copper Mine; subsequently abandoned
1955 May 5 - Claim transferred to Jense McCormick
1959 Feb - Several additional claims made in area by McCormick

Pine Springs Mining District
Ridenour Mine
Minerals noted: *azurite, malachite, chalcocite, chrysocolla, bornite, yellow tyuyamanite, green volborthite, vanadium oxide*
1880 Mar 6 - "Grand Canyon" copper and silver mine claimed by William B. Ridenour, S. Crozier, John Tillman, William H. Hardy, and Charles Spencer
1882 Apr 19 - Renamed Ridenour Mine
1890-1891 - "Copper King" shipped 95 tons of copper
Before and during World War I - Worked intermittently; most activity 1916-1918, when stillowned by Ridenour; ca. 1,000

tons of copper shipped
1920 - Abandoned
1953 - Said to be under control of Hualapai Tribal Council
1955 - Sawyer Exploration Company erroneously implied ownership
1960 May 21 - Clyde Hutcheson signed 10-year lease from Hualapai Tribal Council
1963 or 1964 - Lease canceled by Hutcheson
1976 - Mine leased and drilled by Western Nuclear
1978 - Lease dropped

Lime Ridge prospect (in Hermit Shale)
1892 Oct 28 - Located by William B. Ridenour and J.W. Porter; no activity

National Canyon and Laguna claims (breccia pipes in Esplanade Sandstone and stratiform deposits in Kaibab Formation)
Elements noted: copper, manganese, gold and silver
1891 - Many prospects claimed in National and Moho (Mohawk) Canyons; most by J.A. Healy, S. Gilroy, and J.J. Phillips
1919 Nov 1 - National 1, 2, 3, and 4 claims located by A.M. Boss and Mack Tukespeta in "Walapai Indian Reservation Mining District"
1919 Nov - Lagoon nos. 1, 2, 3, 4, 5, 6, 7, and 8 claims located by Charles H. Dunning and T.E. Carter
1929 - 183 tons of ore shipped at Pica, Arizona, probably from the National Canyon claims
1940s-1950s - Manganese deposits worked south of National Canyon
1983-1987 - Seven small copper prospects rediscovered on Aubrey Ridge and Prospect Valley

Aubrey Cliffs manganese prospect (stratiform breccia deposit)
1940 - Deposit discovered by Don C. Adams and W.J.E. Woody; state permit later lapsed, no production
1953 - E.H. Johnson and R.E. Hayden obtained lease, 37 tons of ore shipped
1953 - Subleased to Bosley Mining Company, 275 tons shipped
1954 Aug - Small quantities of ore produced

Lost Turquoise Mine
Elements noted: copper, uranium
1983 - Rumored mine rediscovered by Mike Bertoldi, Karen Wenrich, and Hoyt Sutphin

Havasu Canyon Mining District
Bridal Veil Mines (including breccia pipes)
Elements noted: silver, lead, zinc, vanadium
1873 June - "Moqui Quartz" claim located by Charles Spencer
Late 1870s - Activity increased
1879 Dec 6 - First claim to be filed in new Yavapai County: Beckman claim, by Daniel W. Mooney, W.C. Beckman, H.J. Young, C.M. Marshall, and Alphonse Humphreys
1881 Feb 9 - "Supai Claim" re-recorded after forfeiture, by W.W. Jones and Matthew Humphreys; named Coconino Silver Mine
1881 Feb 27 - Claims filed by C. Cohen, W.H. Smith, and N. Ellis
1883 - Matthew Humphreys took over Beckman claims

1883 - Claims filed in upper Carbonate Canyon
1885 Sept 23 - Placer claims filed by Phillip McDonnell and Cornelius Clearea
1890 - Claims filed by D.H. Dillon, S.J. Sullivan, John Reese, A.G. Oliver, and Ney Strickland
1892 Sept 7, and - Claims filed by S.J. Sullivan, Paul Dillon, J.R. Dillon,
1893 Jan 1 - S. Morrison, W. Hull, Ney Strickland, and others
1897 Apr 17 - Newspaper reports that J.C. Brown, M. Page Minor, and C.H. Randsburg struck rich gold in "Sullivan's Chimney"—the pillar of gold in the "Randsburg Mining District"
Early 1900s - W.I. Johnson acquired claims, forming Northern Arizona Lead and Zinc Mining Company
Ca. 1902 - Grand Canyon Gold and Platinum Company worked two tunnels at contact between Gateway and Havasu Members of Redwall Limestone
1921 Jan - C.A. Heberlein purchased Johnson claims
World War II - Havasu Lead and Zinc Mining Company attempted to produce lead and vanadium from old Supai claims
1975 - Ownership of all lands and mines on Havasupai Indian Reservation reverted to Havasupai

Hacks Canyon Mining District
Kanab Creek area
1872 Feb - "Kanab gold rush" begins
1872 June - "Kanab gold rush" ends

Hacks Canyon Mine (breccia pipes)
Minerals noted: torbernite, malachite, azurite
Ca. 1890 - Ore first discovered
Ca. 1910 - Road built to mine then called Spotted Bull
Pre-1936 - Various claims in Hacks Canyon
1937-1946 - 14 patented claims comprising Hacks Canyon Mine acquired by Aldus F. "Blondie" Jensen, Clair Pierson, Ray Pointer, and G.C. Harwood:
1937 Sept 24 - Copper Lady
1941 Mar 1 - Hacks
1941 Apr 12 - Hacks No. 2
1941 Sept 5 - Hacks Nos. 3, 4; Mystery
1943 Mar 16 - Hacks No. 5
1946 Feb 26 - Copper Lady Nos. 1-5; Mystery Nos. 2, 4.
1944-1945 - Canyon Copper Company produced some copper
1945 - Uranium discovered in dump spoils
1948-1953 - Copper and uranium produced
1951 July 13 - Hacks Mining Company formed
1952 Dec - Mine leased to California Tungsten Corporation; later merged with Consolidated Uranium Company
1953 Spring - Lease forfeited
1953 May - Option on property taken by Vanadium Corporation of America
1953 July - Option dropped
1953 Sept 7 - LaSalle Mining Company leased claims
1954 Apr - Property subleased to Rainbow Uranium
1954 Aug - Rainbow Uranium reorganized into Uranium; property subsequently subleased to a group of miners; later lawsuit resolved ownership, returning mine to Hacks Mining Company; mine idled
1957 Sept 5 - 90-day lease to Rare Metals Corporation of America

1958 June - Mine leased to Western Gold and Uranium
Early 1959 - Lease dropped
1959 Summer - Mine leased to Atkinson Exploration Company; returned to Western Gold
1963 Sept 5 - Mine leased to A and B Mining Company; later idled
1972 Dec 1 - Mine acquired by Western Nuclear
1977 July - Exploratory drilling nearby
1979 - Hack 2 pipe drilled
1980 July to 1987 June - Hack 2 Mine worked to depletion
1980 Aug - Hack 3 pipe discovered
1981 Jan to 1984 Aug - Hack 1 Mine worked to depletion
1982 Nov to 1987 May - Hack 3 Mine worked to depletion
1988 Apr - Hack 1, 2, and 3 reclaimed
Late 1988 - Mining ended

Thunder River area (contact metamorphism)
Mineral noted: asbestos
No known claims

Grand Canyon Mining District

Bass mines (contact metamorphism)
Elements and minerals noted: asbestos (chrysotile), bornite, chalcopyrite, chalcocite, galena, silver, cuprite
1890s - Copper discovered in Copper Canyon
Post-1890 - Asbestos claims made by Captain Bass in Hakatai Canyon; later copper claims at Swamp Point, on North Rim
1894 - Cable crossing built at Mile 108
1920 Winter - Mine leased to E.L. Quist; cable built at Mile 112
Early 1926 - All Bass properties acquired by Santa Fe Land Development Company

Boucher Mine (secondary replacement in shear zone)
Mineral noted: malachite
1891 - Louis Boucher settled at Dripping Springs, Hermit Basin; later developed small copper mine

Orphan Lode Mine (breccia pipe)
Minerals noted: copper and uranium minerals
1891 - Daniel (John) L. Hogan and Henry Ward discover copper deposit
1893 Feb 8 - Claim made to "the Orphan"
1905 Jan 18-20 - Patent granted by Theodore Roosevelt
1906 Mar 23 - Claim patented with Hogan and Charles J. Babbitt as co-owners
1912 Apr 18 - Babbitt quit-claimed interest to Hogan
1946 Aug 1 - Bertha Madeleine Jacobs bought claim from Hogan
1951 - Amateur prospectors discovered anomalous radioactivity in area
1951 Apr 22 - Harry C. Granger, U.S. Geological Survey, confirmed presence of uranium
1953 Sept - Arthur R. Still, representing Golden Crown Mining Company, leased mine from Jacobs
1955 Dec 14 - Golden Crown announced discovery of uranium to stockholders
1956 - Golden Crown acquired ownership of mine
1956 Apr 25 - First ore shipment, to Atomic Energy Commission buying station at Tuba City, Arizona

1957 July - Western Gold and Uranium, parent company of Golden Crown, acquired full control of Golden Crown
1962 May 28 - Congressional legislation signed by President John F. Kennedy, bringing ownership to mine to National Park Service in May 1987
1966 - Geo Space Corporation acquired ownership of mine; became part of merged corporation, Westec Corporation
1966 Aug 25 - Westec declared bankruptcy
1967 Aug - Westec sold mine to Cotter Corporation
1967 Sept - Cotter Corporation reopened mine and shipped ore to Cañon City, Colorado
1969 Apr 25 - Last ore shipment made
1981 Feb 18 - John R. Siebold and Elling Halvorson acquired ownership of mine
1987 May 28 - Mine ownership reverted to National Park Service
Present - Plans for demolition of surface structures and permanent sealing of mine

Bright Angel Trail claims (non-producing)
1890 Spring - Ralph Cameron and associates staked claims within Grand Canyon and on South Rim
Ca. 1905 - Cameron made more claims in Bright Angel Trail area, up to 39 in one year totaling ca. 13,000 acres
1906 - Berry's franchise to operate Bright Angel Trail as a toll road expired
Early 1907 - 28 claims located on Bright Angel Trail below Devil's Corkscrew
1912 - Philadelphia syndicate took over seven of Cameron's placer claims near Indian Garden; Devil's Corkscrew claims bonded to New York investors
1916 - Santa Fe Land Company acquired key Cameron claims from John Daniell, acting for Cameron, and partner C. Frank Doebler
1923 - Suits filed against Cameron
1928 - Title to Bright Angel Trail passed to National Park Service

Grandview and Last Chance Mines (breccia pipe)
Minerals noted: cyanotrichite, brochantite, chalcoalumite, langite, barite, devilline, chalcanthite, antlerite, gypsum, aurichalcite, azurite, malachite, smithsonite, pyrite, cuprite
1890 Apr 19 - Pete Berry located Grandview and Last Chance mines on Horseshoe Mesa
1895 - Grand Canyon Copper Company formed by Berry together with Ralph and Niles Cameron
1901 Aug 21 - Property sold to Canyon Copper Company
1911 Oct 1 - Berry gave option for sale to William Randolph Hearst
1913 - Hearst bought property
1916 - Mining ended

Hance Mines (contact metamorphism)
Minerals noted: asbestos, chlorite, serpentine, talc
1883-1890 - John Hance prospected in eastern Grand Canyon
1891 Mar - Hance, William H. Ashurst, and John Marshall made claims in Red Canyon area
After 1891 Mar - More claims made in most tributary canyons north of Colorado River, with G.H. McClure, M.M. Fisher, and T.C. Frier joining in some
1894 Jan 1 - Marshall, McClure, Frier, and Ashurst located trail crossing canyon to north side
1895 - J.W. Thurber and H.H. Tolfree bought Hance's hotel and trail
1901 - 16 claims patented, later bonded to the Hance Asbestos

Company, and subsequently bought by George E. Hills, who appointed John Hance supervisor

1904 - Hance removed as supervisor, mines later sold for taxes

Present - Last of Hance mines still in private hands (Hearst family); negotiations for reversion to National Park Service

Nankoweap Canyon area

1891 - Pierce (gold placer) claims made south of Nankoweap Canyon, by Charles Gresswell, Robert Baumgarth, and others

1901 Oct - Robert Baumgarth, J.D. Reese, and Murray Averett located several claims on north side of Nankoweap Butte
-Copper Mountain Claim made in Nankoweap Canyon

Francis Mining District

Anita Mine (stratiform breccia zone)

Minerals noted: chalcopyrite, malachite, azurite, limonite

Pre-1890 - Discovered by Buckey O'Neill

1897 Jan 25 - General W. O'Neill began first work on claims near "Copperopolis"

1897 July - Mr. Nellis and Gerald M. Fennel located several copper claims near Anita and Red Butte

Ca. 1898 - Mine reopened

1912 - Ownership passed to William Nesmith

1914 Jan 1 - William H. and Grace Lockridge filed claim to "Nellis Mine"

Rowe Well Claim (water)

1890 - Sanford A. Rowe established campground at Rowe Well; made "claims" to protect water rights

1956 - Claims acquired by National Park Service

Warm Springs Mining District

Apex Copper properties (stratiform breccia zone)

Elements and minerals noted: malachite, azurite impregnating brecciated zone; earthy cuprite and chalcopyrite; silver and gold

1885 Aug - Andrew F. Greyerson located small copper mine near head of Warm Springs Canyon
-Claims made by Joseph Meeks, Brigham L. Young, J.A. Huntington, and George Adams

1891 Nov 17 - Claims filed by J.S. Best, Harry McDonald, J. Dodson, R.M. Kennedy, and Elmer Kane

1900 - Aquilla Nebeker and V.M. Ryan worked copper mines in area

1900-1902 - Petoskey Mining Company worked five claims in Ryan area

1901 - Coconino Copper Company located on claims later known as the Mackin group

1902-1916 - Experimental processing and production of copper

1907 - Ryan mines ceased operations

1909 - Frank Wynn and Nephi Jennings found uranium specimens on Kaibab Plateau (used in hardening steel and gun metal)

1928 - Saint Anthony Copper Company organized; production of Mackin group of claims; later, title obtained by Los Angeles Exploration and Metal Company Ltd.

1939 - United States Metals Corporation acquired claims; mines reopened under leases

1939-1943 - Production small and intermittent

1943 - Apex Mining Company leased property

1944 - V.M. Ryan and S.B. Atherly subleased Petoskey and Mackin groups of claims

Little Colorado River Mining District

Vicinity of Little Colorado-Colorado River confluence

1880 Apr 3 - Seth B. Tanner, W.H. French, and others formed the "Little Colorado River Mining District" (not the same as the one described herein)

1890 Feb - Ben Beamer began prospecting for copper in eastern Grand Canyon; established residence at confluence of Little Colorado and Colorado Rivers

1891 - Ed Gale, Ralph Cameron, and W.G. Steward made (placer gold?) claims 1.5 miles east of confluence of Little Colorado and Colorado Rivers

1929 July 4 - Mining claim made by note near Big Canyon and Salt Trail Canyon; note found in 1968

McCormick Mine and Tanner Claims

1877 - Dan O'Leary and Charles Spencer claimed silver mine at mouth of Palisades Creek

1889 - Seth B. Tanner, Fred Bunker, and Lewis Bedlias improved old trail from Desert View (Tanner Trail or Bunker Trail)

1890 - Seth Tanner made copper claims in eastern Grand Canyon

1890 - Seth Tanner located several claims between Red Canyon and Little Colorado River

1903 May 27 - George McCormick took over Tanner's mine at Palisades Creek, calling it Copper Blossom Mine; associates were Fred Hensing, George Bailey, A.T. Switzer, and E.O. McCoy

Mines in Chuar and Basalt Canyon areas

1890 Jan 21 - Stanton Expedition meets prospector Felix Lantier on "Furnace Flats"

1890 May - Morning Star claim discovered in Lava Canyon
-Various claims worked by J.N. Hughes, E.A. Reynolds, Al Peterson, Nels Johnson, and Al Huntington

1904 Jan 4 - George Bailey claimed the Copper Grant

1914 Jan 4 - Copper Grant claim refiled

1915 Jan 1 - Marshall Lazure Group Mine located by Charles E. Lazure, J.W.B. Calhoun, and Elliott Hodgen in Basalt Canyon

Lees Ferry Area (undesignated mining district)

Elements noted: gold, rhenium

1879 - Paria River area is a center for gold prospectors

1897 - Julius F. Stone and Robert E. Stanton formed Hoskininni Company to exploit placer gold deposits in Glen Canyon, Utah

1902 - Hoskininni Company bankrupted

1910 - Charles H. Spencer arrived at Lees Ferry; set up hydraulic hose system to extract placer gold from shale

1911 Spring - First "runs" of Spencer's shale washing system; clogging of system due to rhenium interaction with mercury (unknown at the time)

1912 Feb - First voyage of the steamboat Charles H. Spencer

1913 - Placer operation ended

1960s - Spencer attempted to extract rhenium from shales at Paria

APPENDIX 2

Minerals Occurring in Mines and Placer Deposits in the Grand Canyon Region

Mineral	Chemical Formula	Word Formula	Ore Group (Metal)

Native Elements[A]

gold	Au		
silver	Ag		
copper	Cu		
rhenium	Re		

The following elements are indicated by spectrographic analysis of rock samples; the minerals from which they were derived are not known (Kofford, 1969):

cadmium	Cd		
columbium (?)	Cb		
gallium	Ga		
germanium	Ge		
tantalum	Ta		
tin	Sn		
titanium	Ti		
yttrium	Y		

Sulfides

Mineral	Chemical Formula	Word Formula	Ore Group (Metal)
acanthite	Ag_2S	silver sulfide	silver
arsenopyrite	FeAsS	iron arsenic sulfide	iron
bornite	Cu_5FeS_4	copper iron sulfide	copper
bravoite	$(Ni,Fe)S_2$		nickel
chalcocite	Cu_2S	copper sulfide	copper
chalcopyrite	$CuFeS_2$	copper iron sulfide	copper
cinnabar	HgS	mercury sulfide	mercury
covellite	CuS	copper sulfide	copper
digenite	Cu_9S_5		copper
djurleite	$Cu_{31}S_{16}$		copper
galena	PbS	lead sulfide	lead
gersdorffite	NiAsS		nickel
cobalt gersdorffite	$(Ni,Co)AsS$		nickel
lautite	CuAsS	copper arsenide-sulfide	copper
linnaeite	$Co^{+2}Co_2^{+3}S_4$	sulfide of cobalt	cobalt
marcasite	FeS2	iron sulfide	iron
metacinnabar	HgS	mercury sulfide	mercury
millerite	NiS	nickel sulfide	nickel
molybdenite	MoS2	molybdenum sulfide	molybdenum
naumannite	$(Ag_2)Se$	silver selenide	silver
nickeline (niccolite)	NiAs	nickel arsenide	nickel
orpiment, realgar	As_2S_3	arsenic sulfide	arsenic
pararammelsbergite	$NiAs_2$	nickel diarsenide	nickel
polydymite	$NiNi_2S_4$		nickel
pyrite	FeS_2	iron sulfide	iron
rammelsbergite	$NiAs_2$	nickel diarsenide	nickel
siegenite	$(Ni,Co)_3S_4$	cobalt nickel sulfide	nickel
skutterudite	$CoAs_{2-3}$	cobalt arsenide	cobalt
sphalerite	ZnS	zinc sulfide	zinc
stibnite	Sb_2S_3	antimony sulfide	antimony
vaesite	NiS_2	nickel sulfide	nickel
villamaninite	$(Cu,Ni,Co,Fe)S_2$		
violarite	$Fe^{2+}Ni_2^{+3}S_4$		nickel

Sulfo-salts

enargite	Cu_3AsS_4	copper arsenic sulfide	copper
halite	$NaCl$	sodium chloride (salt)	salt
proustite	Ag_3AsS_3	silver arsenic sulfide	silver
tennantite	$(Cu,Fe)_{12}As_4S_{13}$	copper iron arsenic sulfide	copper
tetrahedrite	$(Cu,Fe)_{12}Sb_4S_{13}$	copper iron antimony sulfide	copper

Haloids

fluorite	CaF_2	calcium fluoride

Oxides

bindheimite (?)	$Pb_2Sb_2O_6(O,OH)$		antimony
cuprite	Cu_2O	cuprous oxide	copper
goethite	$\alpha\text{-}Fe^{+3}O(OH)$	iron oxide hydroxide	iron
hematite	$\alpha\text{-}Fe_2O_3$	iron oxide	iron
ilsemannite	$Mo_3O_8 \cdot nH_2O(?)$		molybdenum
limonite	$2Fe_2O_3 \cdot 3H_2O$	[a hydrous iron oxide]	iron
pyrolusite	MnO_2	manganese oxide	manganese
tenorite	CuO	cupric oxide	copper
todorokite	$(Mn^{+2},Ca,Mg)Mn_4^{+4}O_7 \cdot H_2O$		

Carbonates

ankerite	$Ca(Fe^{+2},Mg,Mn)(CO_3)_2$	calcium iron magnesium carbonate	
aragonite	$CaCO_3$	calcium carbonate	
aurichalcite	$(Zn,Cu)_5(CO_3)_2(OH)_6$	zinc copper carbonate hydroxide	zinc
azurite	$Cu_3(CO_3)_2(OH)_2$	copper carbonate hydroxide	copper
bayleyite	$Mg_2(UO_2)(CO_3)_3 \cdot 18H_2O$		uranium
calcite	$CaCO_3$	calcium carbonate	
cerussite	$PbCO_3$	lead carbonate	lead
dolomite	$CaMg(CO_3)_2$	calcium magnesium carbonate	magnesium
malachite	$Cu_2(CO)_3(OH)_2$	copper carbonate hydroxide	copper
rhodochrosite	$MnCo_3$	manganese carbonate	manganese
schroeckingerite	$NaCa_3(UO_2)(CO_3)_3(SO_4)F \cdot 10H_2O$		uranium
siderite	$Fe^{+2}CO_3$	iron carbonate	iron
smithsonite	$ZnCo_3$	zinc carbonate	zinc

Silicates

biotite	$K(Mg,Fe^{+2})_3(Al,Fe^{+3})Si_3O_{10}(OH,F)_2$	potassium magnesium iron aluminosilicate	hydroxide
celadonite	$K(Mg,Fe^{+2})(Fe^{+3},Al)Si_4O_{10}(OH)_2$		
chalcedony	SiO_2		
chert	SiO_2	[a rock type derived from chemical replacement]	
chlorite	chlorite group formula:		
	$\varepsilon_{4\text{-}6}\zeta_4O_{10}(OH,O)_8;$		
	ε = Al, Fe^{+2}, Fe^{+3}, Li, Mg, Mn^{+2}, Ni		
	ζ = Al, B, Fe^{+3}, Si		
chrysocolla	$(Cu,Al)_2H_2Si_2O_5 \cdot nH2O$	copper silicate hydroxide	copper
coffinite	$U(SiO_4)_{1\text{-}x}(OH)_{4x}$		uranium
hemimorphite	$Zn_4Si_2O_7(OH)_2 \cdot H_2O$	zinc silicate hydroxide hydrate	zinc
illite	illite group formula:		
	$(K,H_3O)(Al,Mg,Fe)_2(Si,Al)_4O_{10}[(OH)_2,H_2O]$		
kaolinite	$Al_2Si_2O_5(OH)_4$		
leonhardite (starkeyite)	$MgSo_4 \cdot 4H_2O$		
microcline	$KAlSi_3O_8$	potassium aluminosilicate	
plagioclase	feldspar series		
	general formula: $(Na,Ca)Al(Al,Si)S_2O_8$		

quartz	SiO_2	silicon oxide	
roscoelite	$K(V,Al,Mg)_2AlSi_3O_{10}(OH)_2$		vanadium
sericite (muscovite)	$KAl_2(Si_3Al)O_{10}(OH,F)_2$	potassium aluminosilicate hydroxide	
tourmaline	tourmaline group formula: $\epsilon\zeta_3\eta_6(BO_3)_3Si_6O_{18}(O,OH,F)_4$; ϵ = Ca, K, Na ζ = Al, Fe^{+2}, Fe^{+3}, Li, Mg, Mn^{+2} η = Al, Cr^{+3}, Fe^{+3}, V^{+3}		
uranophane	$(H_3O)_2Ca(UO_2)_2(SiO_4)_2 \cdot 3H_2O$	calcium uranyl phosphate hydrate	uranium
zaratite (?)	$Ni_3(CO_3)(OH)_4 \cdot 4H_2O$		nickel
zircon	$ZrSiO_4$	zirconium silicate	

Phosphates

metatorbernite	$Cu(UO_2)_2(PO_4)_2 \cdot 8H_2O$	copper uranyl phosphate hydrate	uranium

Uranates

uraninite	UO_2 (oxidizing to U_3O_8)		uranium

Arsenates

adamite	$Zn_2(AsO_4)(OH)$	zinc arsenate hydroxide	zinc
annabergite (?)	$Ni_3(AsO_4)_2 \cdot 8H_2O$		nickel
conichalcite	$CaCu(AsO_4)(OH)$	calcium copper arsenate hydroxide	copper
erythrite	$Co_3(AsO_4)_2 \cdot 8H_2O$	cobalt arsenate hydrate	cobalt
metazeunerite	$Cu(UO_2)_2(AsO_4)_2 \cdot 8H_2O$		uranium
olivenite	$Cu_2AsO_4(OH)$	copper arsenate hydroxide	copper
scorodite	$Fe^{+3}AsO_4 \cdot 2H_2O$	iron arsenate hydrate	iron
talmessite	$Ca_2Mg(AsO_4)_2 \cdot 2H_2O$		
uranospinite	$Ca(UO_2)_2(AsO_4)_2 \cdot 10H_2O$		uranium
zeunerite	$Cu(UO_2)_2(AsO_4)_2 \cdot 10\text{-}16H_2O$	copper uranyl arsenate hydrate	uranium

Vanadates

calciovolborthite	$CaCu(VO_4)(OH)$		vanadium
carnotite[B]	$K_2(UO_2)_2V_2O_8 \cdot 3H_2O$	potassium uranyl vanadate hydrate	uranium
hewettite	$CaV_6O_{16} \cdot 9H_2O$		vanadium
metatyuyamunite	$Ca(UO_2)_2V_2O_8 \cdot 3H_2O$		uranium
tyuyamunite	$Ca(UO_2)_2V_2O_8 \cdot 5\text{-}8H_2O$	calcium uranyl vanadate hydrate	uranium
vesignieite	$BaCu_3(VO_4)_2(OH)_2$		vanadium
volborthite	$Cu_3(VO_4)_2 \cdot 3H_2O$		vanadium

Sulfates

anglesite	$PbSO_4$	lead sulfate	lead
anhydrite	$CaSO_4$	calcium sulfate	
antlerite	$Cu_3(SO_4)(OH)_4$	copper sulfate hydroxide	copper
barite	$BaSO_4$	barium sulfate	barium
bieberite	$CoSO_4 \cdot 7H_2O$		cobalt
brochantite	$Cu_4(SO_4)(OH)_6$	copper sulfate hydroxide	copper
chalcanthite	$CuSO_4 \cdot 5H_2O$		copper
chalcoalumite	$CuAl_4(SO_4)(OH)_{12} \cdot 3H_2O$		copper
coquimbite	$Fe_2^{+3}(SO_4)_3 \cdot 9H_2O$		iron
cyanotrichite	$Cu_4Al_2(SO_4)(OH)_{12} \cdot 2H_2O$	copper aluminum sulfate hydroxide hydrate	copper
devilline	$CaCu_4(SO_4)_2(OH)_6 \cdot 3H_2O$		copper
gypsum	$CaSO_4 \cdot 2H_2O$	calcium sulfate hydrate	
hexahydrite	$MgSO_4 \cdot 6H_2O$		
jarosite	$KFe_3^{+3}(SO_4)_2(OH)_6$	potassium iron sulfate hydroxide	
johannite	$Cu(UO_2)(SO_4)_2(OH_2) \cdot 8H_2O$		uranium
langite	$Cu_4(SO_4)(OH)_6 \cdot 2H_2O$	hydrous sulfate of copper	copper

melanterite	$Fe^{+2}SO_4 \cdot 7H_2O$		iron
siderotil	$Fe^{+2}SO_4 \cdot 5H_2O$		iron
uranopilite	$(UO_2)_6(SO_4)(OH)_{10} \cdot 12H_2O$		uranium
zippeite	$K_4(UO_2)_6(SO_4)_3(OH)_{10} \cdot 4H_2O$		uranium

Molybdates ————————————————————————————————

| powellite | $CaMoO_4$ | calcium molybdate | molybdenum |
| wulfenite | $PbMoO_4$ | lead molybdate | lead |

Native American Trade Items ————————————————————

| turquoise | $CuAl_6(PO_4)_4(OH)_8 \cdot 5H_2O$ | copper aluminum phosphate hydroxide | hydrate |
| obsidian | [a black, glassy, compact form of rhyolite; a dense rock composed essentially of alkali feldspar and quartz] | | |

[A] In addition, platinum (Pt) has been reported by early miners, but its occurrence in the Grand Canyon is uncorroborated.
[B] Yellow carnotite is also reported here from an X-ray diffraction analysis by Spamer, in 1977, of a sample from the Tapeats Sandstone in Bright Angel Canyon. The occurrence of carnotite there is a secondary emplacement. It may have been redeposited at depth in the Bright Angel fault zone by groundwater at a time when the Mesozoic uraniferous strata once overlay the Grand Canyon region, thus it may be contemporaneous with breccia pipe mineralization.

Most mineral occurrences are within breccia pipes, but not all pipes may be mineralized. Many minerals in breccia pipes have been identified only in laboratory analytical data. Primary references for this table are Gornitz (1969), Kofford (1969), and Wenrich & Sutphin (1987, 1988), with additional data from Wenrich & Hlava (1993). Most chemical formulae are as given by Fleischer (1987); others were obtained from mineralogical textbooks. Word formulas are taken from White's (1987) list in Fleischer.

Table 1. Production Data for the Copper Mountain Mine (1934-1953)[1]

Year	Ore shipped (tns)	Copper (oz)	Lead (lbs)	Zinc (lbs)	Gold (lbs)	Silver (oz)
1934	16	5779			0.4	81
1937	6	1750				20
1949	195	12,625	8,367	37,994	163	1,728
1950	151	94,000			2	476
1951	23	4,936				47
1952	21	4,774				28
1953	133	3,814				28
Total	545	127,681	8,367	37,994	165.4	2,411

[1]From Foord et al. (1978, table 2)

Table 2. Assay of the Bridal Veil Mine

Claim	Silver (oz/tn)	Lead (%)	Zinc (%)	Vanadium Oxide (%)
Cataract		0.60	1.70	0
Iron Mask	0.20	0.15		
Seligman	3.4	11.70		0.41
Road		0.20	0	
Bridal Veil	15.7	67.5		1.72
Blende	5	0.15	26.35	
Home	1.20	2.40		0
Pyramid	6.7	28.95		0
Elenia	0.8	0.25	0.70	0.25

Table 3. Wells Drilled for Oil and Natural Gas in Northwestern Arizona (after Giardina, 1979:43)

ID[1]	Location[2]	Operator, Number, and Lease	Original Completion Date	Total Depth (feet)	Stratigraphic Unit or System at Total Depth
Coconino County					
275	39N 02E 32 NE NE	Underwood 1-32 Jacob Lake Unit (Federal)	May 1964	3868	Bright Angel (Cambrian)
Mohave County					
43	38N 05W 31 NW SE	Western Drilling–Valen Oil and Gas 1 (Federal)	May 1958	4666	Bright Angel Shale (Cambrian)
41	38N 07W 17 SW SW	Roger A Fields 1 (Federal)	June 1957	460	Toroweap(?) Formation (Permian)
53	38N 07W 17 SW SW	Roger A Fields 1-X (Federal)	Apr 1958	1780	Permian
502	38N 07W 29 NW NE	James J Harris 1 (Federal)	July 1970	1115	Hermit Shale (Permian)
42	39N 06W 14 SW NW	Paul Poteet & Tony Lyons 1 (Federal)	Aug 1957	2303	Coconino Sandstone (Permian)
56	39N 06W 35 NE SW	Tony Lyons 1 (Federal)	Sept 1958	1820	Toroweap Formation (Permian)
347	39N 07W 02 NE SE	Skelly Oil 1-A (Federal)	May 1966	4031	Mooney Falls Member of Redwall Limestone (Mississippian)
114	39N 13W 35 SE SW	Tennessee Gas Transmission 1 Schreiber (Federal)	May 1960	4015	Mississippian
40	40N 06W 12 NW SW	T W George 1 (Federal)	June 1957	2202	Kaibab Formation (Permian)
33	40N 08W 28 SE SW	Falcon Seaboard Drilling–Valen Oil and Minerals 1 Antelope (Federal)	Aug 1956	3753	Mississippian
677	40N 09W 18 NE SW	Pyramid Oil 2 (Federal)	Nov 1977	4509	Devonian
8-19	41N 06W 16 SE	Cane Bed 1 (State)	1931	542	
8-20	41N 08W 18 NE NW	Antelope Petroleum 1 Morris (Federal)	1932	1522	
676	41N 09W 28 NW SE	Pyramid Oil 1 (Federal)	Dec 1977	4150	Mississippian
8-21	41N 15W 29 SW SE	Virgin Oil 4 (State)	1918	2600	
8-22	42N 08W 31 SW SW	Arizona and Utah Consolidated Oil 1 (State)	1909	936	
8-23	42N 15W 32	Virgin Oil and Mines 1 (State)	1931	1405	
8-24	42N 15W 32	Virgin Oil and Mines 6 (State)	1931	545	

[1]Arizona Oil and Gas Conservation Commission identification number.

[2]Township and Range (N, W), Section, and nested quadrants within section.

INDEX

MAP 1 - Physiographic Map

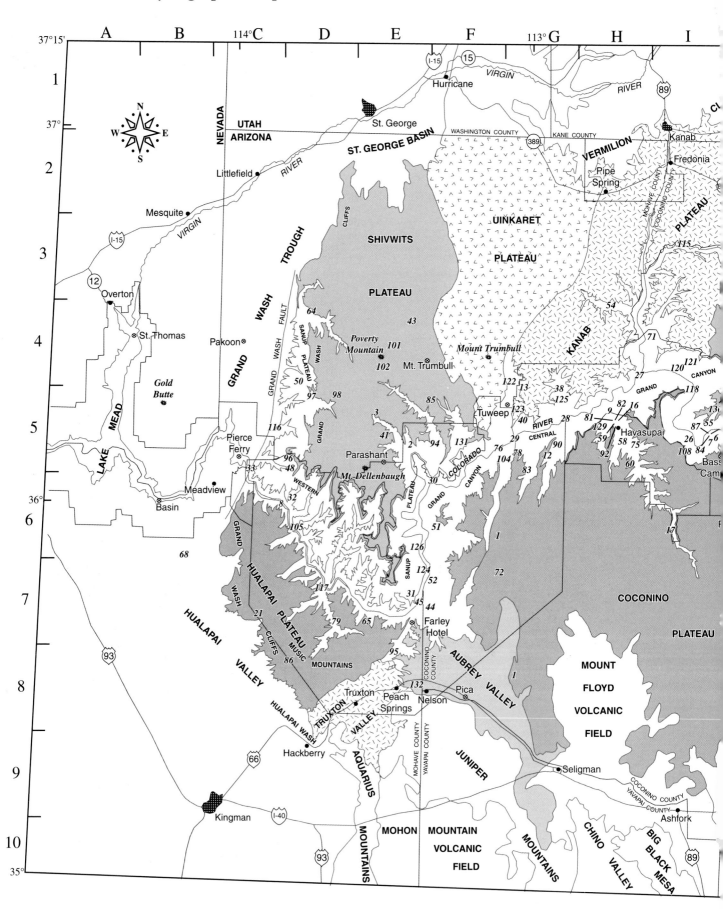

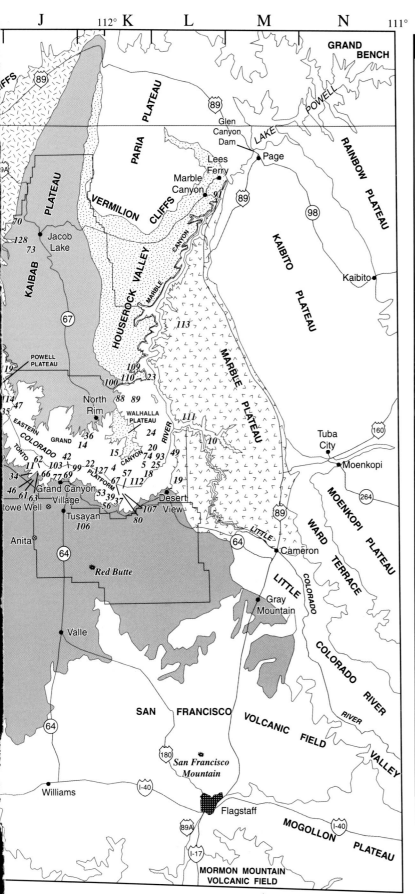

MAP 1

1.	Aubrey Cliffs	6-F, 8-G	67.	Horseshoe Mesa	6-K
2.	Andrus Canyon	5-E	68.	Hualapai Wash	6-B
3.	Andrus Spring	5-E	69.	Indian Garden	6-J
4.	Asbestos Canyon	6-K	70.	Jacob Canyon	3-J
5.	Basalt Canyon	6-K	71.	Kanab Creek Canyon	4-H
6.	Bass Canyon	5-I	72.	Laguna Lake	7-F
7.	Bass Point	5-I	73.	Lambs Lake	3-J
8.	Bat Tower Viewpoint	6-C	74.	Lava Canyon	6-K
9.	Beaver Falls	5-H	75.	Lee Canyon	5-H
10.	Big Canyon	5-L	76.	Lime Ridge	5-F
11.	Boucher Canyon	6-J	77.	Maricopa Point	6-J
12.	Boulder Canyon	5-G	78.	McKee Point	5-G
13.	Brady Canyon	5-G	79.	Milkweed Canyon	7-D
14.	Bright Angel Canyon	5-K	80.	Miners Spring	6-K
15.	Cape Royal	6-K	81.	Mooney Canyon	5-H
16.	Carbonate Canyon	5-H	82.	Mooney Falls	5-H
17.	Cataract Creek	6-I	83.	Mohawk Canyon	5-G
18.	Cardenas Creek	6-K	84.	Mt. Huethawali	5-I
19.	Cedar Mountain	6-L	85.	Mule Canyon	5-E
20.	Chuar Creek	5-L	86.	Music Mountains	8-D
21.	Clay Springs	7-C	87.	Mystic Spring	5-I
22.	Clear Creek	6-K	88.	Nankoweap Canyon	5-K
23.	Colorado River Mile-52	5-L	89.	Nankoweap Butte	5-K
24.	Colorado River Mile-61	5-L	90.	National Canyon	5-G
25.	Colorado River Mile-66	6-L	91.	Navajo Bridge	2-L
26.	Colorado River Mile-116	5-I	92.	Navajo Falls	5-H
27.	Colorado River Mile-144	4-H	93.	Palisades Creek	6-L
28.	Colorado River Mile-166	5-G	94.	Parashant Canyon	5-F
29.	Colorado River Mile-180	5-G	95.	Peach Springs Canyon	8-E
30.	Colorado River Mile-200	6-E	96.	Pierce Canyon	5-D
31.	Colorado River Mile-225	7-E	97.	Pigeon Canyon	5-D
32.	Colorado River Mile-266	6-D	98.	Pigeon Springs	5-D
33.	Colorado River Mile-277	5-C	99.	Pipe Creek	6-K
34.	Columbus (Yuma) Point	6-J	100.	Point Imperial	5-K
35.	Copper Canyon	5-I	101.	Poverty Knoll	4-K
36.	Cottonwood Camp	5-K	102.	Poverty Mountain	4-E
37.	Cottonwood Creek	6-K	103.	Powell Point	6-J
38.	Cottonwood Spring	5-G	104.	Prospect Canyon	5-F
39.	Cottonwood Springs	6-K	105.	Quartermaster Canyon	6-D
40.	Cove Canyon	5-G	106.	Red Butte	6-K
41.	Dansill Canyon	5-E	107.	Red Canyon	6-K
42.	Devil's Corkscrew	6-J	108.	Royal Arch Creek	5-I
43.	Diamond Butte	4-E	109.	Saddle Canyon	4-K
44.	Diamond Creek	7-E	110.	Saddle Mountain	5-K
45.	Diamond Peak	7-E	111.	Salt Trail Canyon	5-L
46.	Dripping Springs	6-J	112.	Seventy-five Mile Canyon	6-K
47.	Flint Creek	5-J	113.	Shinumo Altar	4-L
48.	Fort Garrett	5-D	114.	Shinumo Creek	5-J
49.	Gold Hill	6-L	115.	Snake Gulch	3-I
50.	Grand Gulch Bench	4-D	116.	Snap Canyon	5-C
51.	Granite Park	6-F	117.	Spencer Canyon	7-D
52.	Granite Spring Rapids	7-E	118.	Stone Creek	5-I
53.	Grapevine Canyon	6-K	119.	Swamp Point	5-J
54.	Hacks Canyon	4-H	120.	Tapeats Creek	4-I
55.	Hakatai Canyon	5-I	121.	Thunder River	4-I
56.	Hance Canyon	6-K	122.	Toroweap Canyon	4-G
57.	Hance Rapids	6-K	123.	Toroweap Point	5-G
58.	Havasu Creek	5-H	124.	Trail Canyon	7-E
59.	Havasu Falls	5-H	125.	Tuckup Canyon	5-G
60.	Havasupsi Canyon	5-H	126.	Two-fourteen Mile Canyon	6-E
61.	Hermit Basin	6-J	127.	Vishnu Canyon	6-K
62.	Hermit Camp	6-J	128.	Warm Springs Canyon	3-J
63.	Hermit Canyon	6-J	129.	Watahomigi Point	5-H
64.	Hidden Canyon	4-D	130.	White Creek	5-I
65.	Hindu Canyon	7-E	131.	Whitmore Canyon	5-F
66.	Horn Creek	6-J	132.	Yampai Canyon	8-E

⊗ abandoned settlements   ● small settlements   ▓ large settlements

Map credits: Zenon Valin, Haydee Hampton, and Kim Buchheit